BIOLOGY OF SEAWEEDS

Levels of Organization

by

A. R. O. Chapman, Ph. D.

Dalhousie University
Halifax, Nova Scotia

Illustrations by
Pat Evans and **Terry Collins**

University Park Press
Baltimore

UNIVERSITY PARK PRESS
International Publishers in Science, Medicine, and Education
233 East Redwood Street
Baltimore, Maryland 21202

Typeset by American Graphic Arts Corporation
Manufactured in the United States of America by
Universal Lithographers, Inc.,
and The Maple Press Company

Library of Congress Cataloging in Publication Data

Chapman, A.R.O.
Biology of seaweeds.

Includes bibliographical references and index.
1. Marine algae. 2. Plant communities. I. Title.
QK570.2.C47 589'.39'2 78-22104
ISBN 0-8391-1340-4

To Jan

Contents

Preface

Many Departments of Botany in universities and colleges offer a course of perhaps 20 lectures and half as many laboratories in phycology. Phycology covers the whole of algal biology, a somewhat awesome topic for review in 20 lectures. A teacher is faced with the difficult prospect of selecting a suitable approach. Traditionally this approach has consisted of systematic review of all of the algal phyla or divisions. For a course such as this there are many available texts (for example Bold, 1967; Bold and Wynne, 1978; Prescott, 1968; Scagel et al., 1965). Alternatively the algae may be considered in terms of their functional organization, considering the structure and function of cells, whole organisms, and communities. This approach has been used, for example, by Round (1973) and Venkataraman et al. (1974), who give a minimal systematic treatment. A hybrid approach has been used by Gayral (1975).

The organizational approach is used here. I have put together in a single volume part of a series of lectures given at the third-year level at Dalhousie University. Unlike the other texts quoted above, attention is focused mainly on seaweed algae, and considerable emphasis is placed on ecological considerations. By narrowing the scope in this way, it is possible to point out unifying trends and concepts—something that is always difficult for a phycology teacher.

A basic knowledge of biology is required by the reader of this book. First or second year courses in principles of biology and diversity of organisms should be adequate prerequisites.

A. R. O. Chapman

Acknowledgments

The line drawings in this book were prepared by Pat Evans and Terry Collins. I am very grateful for their skilled assistance. Most of the drawings are adaptations of previous illustrations, and I would like to thank copyright holders and authors for permission to use their work.

I am especially grateful to Dr. J. McLachlan, who very kindly read and criticized the manuscript at very short notice.

Introduction

The concept of levels of organization is an old one. In biology the highest level of organization is that of the community, which is an assemblage of populations of different species. Single species populations form the next lowest level, followed by whole organisms made up of tissues which are in turn made up of cells. Below the cellular level of organization there are the molecular and biochemical levels. In this book the seaweed algae are examined at each of the levels of organization between cell and community. Tissues and organs do not warrant separate treatment because they are not well developed in algae. Thus there are four separate parts to the book, dealing in turn with cells, whole organisms, populations, and communities.

Seaweeds are macroscopic members of the divisions Chlorophyta, Phaeophyta, and Rhodophyta living in the sea. As communities they are easy to recognize. They are plants visible to the naked eye, generally growing attached to solid substrata between and below the tide marks. This is not a watertight diagnosis, because this community also includes the angiosperm (flowering plant) sea grasses, which are not considered here. Seaweeds may also occur in an unattached state, as for example in the Sargasso Sea. However, attached plants form by far the greater proportion of the total seaweed stock of the world.

The divisions Chlorophyta, Phaeophyta, and Rhodophyta are diagnosed in Table 1. Notice that the criteria are cytological and biochemical. This is a reflection of the great diversity of algae at these levels of organization.

The Rhodophyta and Chlorophyta contain unicellular and multicellular microscopic forms. These are not considered here. The Phaeophyta comprises only multicellular forms.

We now have a formal definition of the seaweed algae, and we know where they live. To the layman they may seem rather unpleasant organisms—very slimy, unpleasant to look at and to touch. This is an uneducated opinion. Some of these plants are among the most beautiful in the world, especially the red algae or Rhodophyta. They are also academically and commercially interesting. Such is the interest they generate that a major International Seaweed Symposium is convened every 3 years to present information and ideas on their biology, chemistry, and commercial significance.

Seaweeds are worth studying at the cytological level, for example, because of the diversity of cell wall structures. They give us clues about how the physics of a wall controls the growth of cells. At the whole organism level we see many fundamentally different approaches in the construction of multicellular plant bodies. Of special interest to the developmental biologist is the fact that, by the aggregation of filaments, well-organized plants of 6 m in length may be produced.

Seaweed population biology is in its infancy, but already we see the potential usefulness of unusual life histories in genetic analysis. Higher plants and animals display very little diversity of life history. The theory of popula-

Table 1. Diagnosis of Chlorophyta, Phaeophyta, and Rhodophyta

Division	Pigments and plastids	Storage products	Cell walls	Flagellation
Chlorophyta	Chlorophyll *a*, *b* Xanthophylls Two to six thylakoids per stack	Starch, as in higher plants	Cellulose, hydroxyproline, glycosides, or wall absent may be calcified	One, two to eight Whiplash, apical insertion, equal length
Phaeophyta	Chlorophyll *a*, *c* β-carotene xanthophylls, including fucoxanthin Three thylakoids per stack	Laminarin, mannitol	Cellulose, alginic acid, sulfated polysaccharides	2, unequal length, lateral insertion
Rhodophyta	Chlorophyll *a*, *d* c- and r- phycocyanin, r-phycoerythrin, α- and β-carotene Several xanthophylls, thylakoids not stacked	Floridean starch	Cellulose in most, xylans, pectin, sulfated polysaccharides	None

After Bold (1967).

tion biology is built around the diploid organism. Seaweeds may have independent diploid and haploid stages in the same life history, however, and these may be morphologically similar or dissimilar. This should provide rich material for the theorist and for the experimentalist.

Seaweed communities are providing excellent subjects for a new approach to community structure—the experimental. Ideas on community structure are often based on observation alone. This is not surprising, because in a terrestrial forest the trees may take 100 years to reach full size. This is rather too long for a Ph.D. research project. On the other hand, seaweed forests may grow to 30 m in 6 months. The recent experiments of Paul K. Dayton in the seaweed beds of the western United States are revealing a great deal of the interaction of the components that make up the structure of a community.

The commercial aspect of seaweeds is not considered here since it is well treated by Boney (1965) and V. J. Chapman (1970). They are not as important, obviously, as land crops. However, in certain parts of the world seaweed farming is a major industry.

Evolutionary aspects of algae are not considered here either, although this is certainly an interesting topic. Algae were probably among the first eukaryotic organisms to evolve. As such, the current speculation about the origin of the eukaryotic cell must look at the features of modern day algae. The only direct evidence on phylogeny comes from the fossil record. There have been exciting developments in the palaeontology of algae in recent years. These have been put together in a major review by Schopf (1970).

BIOLOGY OF SEAWEEDS

Part I
Organization in Cells

Seaweeds are eukaryotes, that is, they have membrane-bounded cellular organelles. Within the eukaryote algae as a whole there is an overwhelming diversity of structures. The seaweeds are more uniform, but there is still a great deal of variability, especially in the Chlorophyta. Eukaryotes ranging from algae to man share many cytological features. The generalities of cell structure and function are dealt with in almost all introductory biology texts. Emphasis is placed here on those cytological characteristics that are special to seaweeds or that are best exemplified in seaweeds. Much of the ultrastructural work on algae has been carried out on freshwater and unicellular forms. Although this is a book about seaweeds, reference to other algae may be unavoidable here. However, treatment is restricted to the phyla Chlorophyta, Phaeophyta, and Rhodophyta.

Chapter 1
Cell Structure

CELL WALLS

Light microscopic examination of sections of red, green, and brown algae show an outermost layer that is about 1 μm thick (Figure 1). Hanic and Craigie (1969) term this a "cuticle," although it is obviously not homologous with the cuticle of higher plants. In seaweeds the cuticle appears to be about 80% proteinaceous.

The cell walls surrounding individual cells are primarily of carbohydrate constituents. The carbohydrates are in two forms, a fibrillar form and an amorphous form. The fibrils (called microfibrils) are from 3 nm to 20 nm in diameter and are commonly cellulosic. Cellulose is a polymer of β-1,4-glucose. Some microfibrils are polymers of β-1,4-mannose and β-1,3-xylose. The chemistry of the fibrils does not seem to influence their biophysical properties.

The nonfibrillar polysaccharides in the wall are mucilaginous and of enormously varied and complex chemical composition.

On the basis of microfibrillar organization, Mackie and Preston (1974) have divided algal walls into two types: the *Valonia* type and other wall types. The *Valonia*-type walls are laminated and are found in only a few green algae. Fibrils lie primarily in two orientations at 90° to one another (Figure 2) and the two orientations occur in different lamellae. A third orientation occurs at an intermediate angle. These walls show no differentiation between primary and secondary walls as found in higher plants. The *Valonia* type of construction is often called "crossed-fibrillar."

Most filamentous red, green, and brown algae differ from *Valonia* in having a bipartite wall. During the growth of green algae with this type of construction, a thin wall layer is present and in this the microfibrils are oriented almost transversely, like hoops around the cell periphery. During cell differentiation, a thicker wall layer is deposited and in this the microfibrils are oriented almost longitudinally. This is called a multi-net construction.

In the brown algae the fibrils are usually dispersed at random. A notable exception is found in the long internal hyphal cells of complex forms where the fibrils are oriented parallel to the long axis. In red algae the fibrils are dispersed at random.

Although the microfibrils in the algae vary little in chemical composition and are, for the most part, simple polymers, the mucilages of the amorphous phase are very complex. In the brown algae the major component is alginic acid, which is a block copolymer of 1,4-linked D-mannuronic acid and 1,4-linked L-guluronic acid. The tertiary structure of the molecule is in the form of a helix. The biological functions of alginic acid are mainly of a structural and ion-exchange type. The composition of the alginate block segments (i.e., the arrangement of mannuronic and guluronic residues) determines both cation selectivity and flexibility. According to Mackie and Preston (1974), alginate

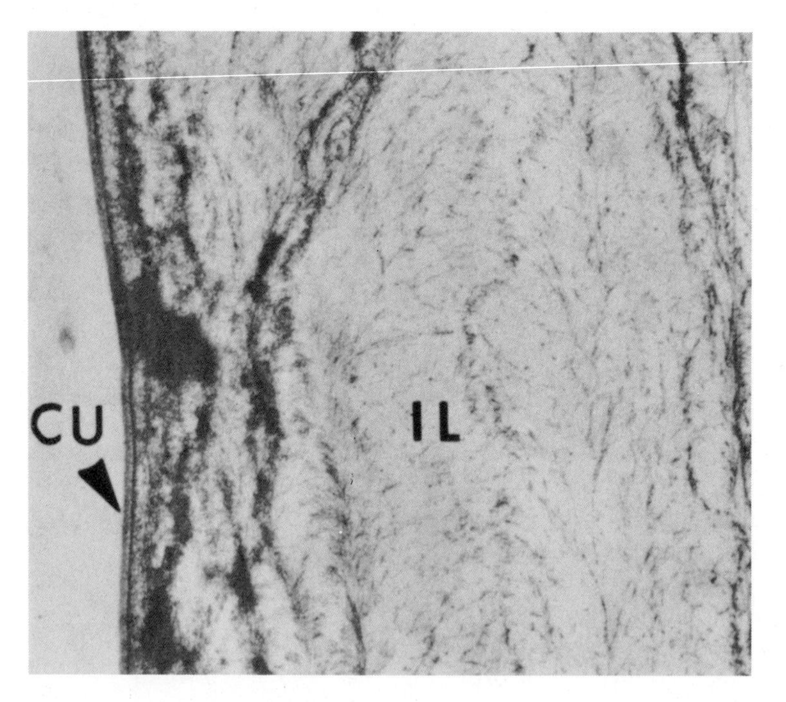

Figure 1. Cross section of the cell wall of *Spongomorpha arcta* showing a discrete cuticle (CU) and an inner thick fibrous layer (IL). × 20,000. From Hanic and Craigie (1969), courtesy of L. A. Hanic and the *Journal of Phycology*.

Figure 2. Inner lamellae of a side cell wall of *Chaetomorpha melagonium* viewed from the outside. × 30,000. From Mackie and Preston (1974). By permission of Blackwell Scientific Publications, Ltd.

structure may determine the flexibility of different tissues in different species according to environment.

In the red algae the wall matrix consists of sulfated polysaccharides, such as the galactans agar and carrageenan. The basic units of the galactan molecule are β-1,3- and α-1,4-linked. Beyond this there are enormous complexity and diversity in the sulfated polysaccharides of the Rhodophyta (see Mackie and Preston, 1974; Percival and McDowell, 1967).

The gelling properties of the matrix polysaccharides of red and brown algae have not been overlooked by commercial interests. An industry exists to extract these substances for use in food and cosmetic enterprises.

The growth of cell walls and their control over morphogenesis has been extensively investigated in the green algae with multi-net fibrillar construction. This aspect of cell wall function is considered in some detail in Chapter 2.

Although some brief mention has been made of the chemistry of walls, it should be realized that this book is not directly concerned with the biochemical level of organization. For a full treatment of this aspect reference should be made to Percival and McDowell (1967), Mackie and Preston (1974), and Preston (1974).

Calcification

In many parts of the world seaweeds with an encrusting coat of calcium carbonate can be found. Most of these plants belong to the red algae, although some green algae are also calcified (e.g., *Halimeda*). In *Halimeda* the $CaCO_3$ as aragonite fills the space between the utricles and is not found in the cell wall. In red algae the crystals are deposited in the matrix of the wall. As with many other wall-building processes, the Golgi seems involved in crystal formation.

FLAGELLA

Flagella are found in the reproductive bodies of green and brown seaweeds. Red algae do not possess flagella, although some kind of amoeboid movement may be evident in spores (Dixon, 1973). Simon-Bichard-Bréaud (1971) has reported flagella in the spermatangial branch of *Bonnemaisonia hamifera*, but Dixon (1973) suspects that this is a consequence of infection by a fungus and its subsequent formation of motile reproductive bodies. The absence of a flagellum in the red algae and its ubiquity in other eukaryotes is strong evidence that this group of seaweeds diverged from the main line of eukaryote evolution at a very early stage. There has been considerable speculation on the phylogenetic relationships between algal groups and also on the origin of land plants from aquatic algae (see Klein and Cronquist, 1967). The fossil record is of little help. The Rhodophyta, Phaeophyta, and Chlorophyta appeared as quite distinct groups as far back as 600 million years ago (Schopf, 1970). The unicellular Chlorophytes may have arisen 1,000 million years ago as the first eukaryotes.

Within the green and brown seaweeds, flagella are arranged in one of three ways, shown in Figure 3. In the isokont (Figure 3A) condition the flagella are identical in length. In most green algae they are inserted apically. In brown algae the flagella are unequal and inserted laterally (Figure 3B). In the green seaweed *Derbesia* the flagella are arranged like a crown around the spore apex. This is called the stephanokont condition (Figure 3C).

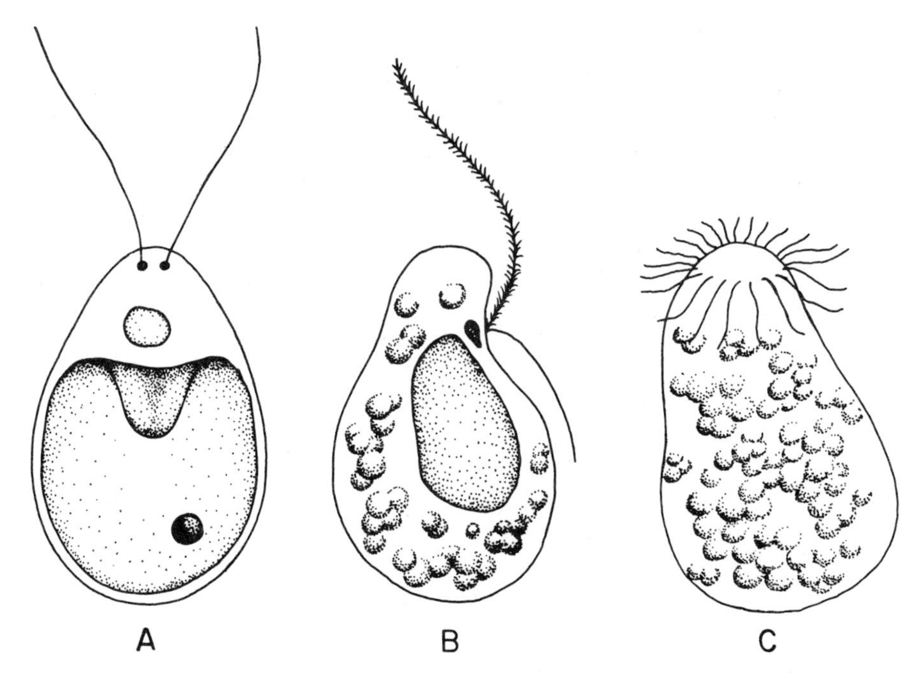

Figure 3. Flagellar arrangements in seaweeds. A, isokont; B, heterokont; C, stephanokont. After Gayral (1975). By permission of Doin, Editeurs.

Most of the variation in the structure of individual flagella is in appendages external to the flagellum plasma membrane. The main section of the free moving part of the flagellum has the classic 9 + 2 configuration, where there are nine peripheral and two central tubules (Figure 4). The microtubular apparatus is surrounded by a granular matrix and covered by an extension of the cellular plasma membrane. In the sperms of the large brown seaweeds, *Fucus*, *Ascophyllum*, and *Pelvetia*, there is a well-developed striate proboscis. This proboscis extends back along the body of the sperm as a strand of fibrils (Manton and Clarke, 1951). Where the flagellum enters the cell a complex transition takes place. The two central tubules terminate at the basal body (Figure 4). The peripheral tubules run into the cell cytoplasm, where their arrangement changes as shown in Figure 4. The ring of tubules is known as a basal body and to this the flagellar roots are attached. The function of these roots—anchorage or stimulus transmission—is undetermined.

In the sperms of *Halidrys* and *Cystoseira* (Manton, 1964) and zoospores of *Scytosiphon* the flagella have a two-limbed root consisting of six to nine fibers running in close association with mitochrondria.

Flagella of the algae may be endowed with superficial appendages. The appendages may be hairs (e.g., *Ascophyllum*) or spines (e.g., *Himanthalia* and *Dictyota*). Flagellar hairs originate from fibril containing vesicles in the cell cytoplasm (Bouck, 1969).

The most interesting question to be asked about flagella is "how do they work?" After more than a hundred years of examination, the question remains essentially unanswered. They may move in one of

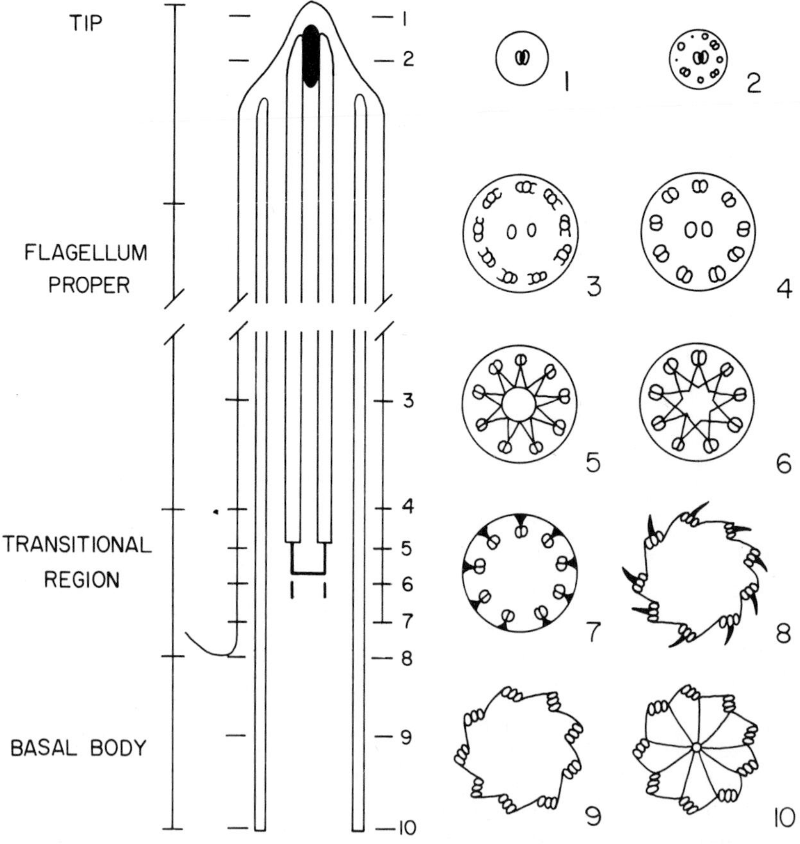

Figure 4. Stylized longitudinal section of the flagellum of *Chlamydomonas*. Ten cross-sections are shown for each of the numbered points on the longitudinal section. After Ringo (1967).

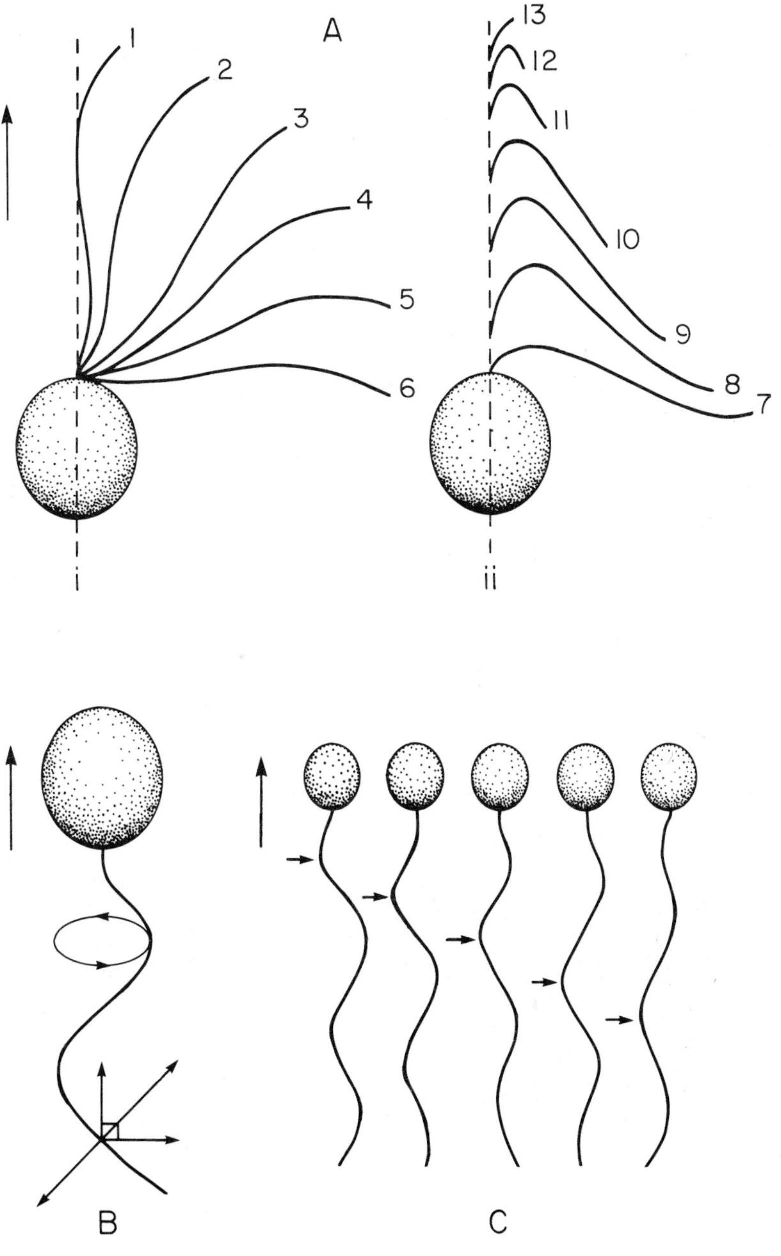

Figure 5. Types of flagellar propulsion. A, pull type showing a) power stroke and b) return stroke; B, propeller type; C, undulatory type. After Nultsch (1974). By permission of Blackwell Scientific Publications, Ltd.

three ways (Figure 5). The best hypothesis is that during movement the contractile fibrils undergo an alternation of contraction and relaxation resulting in unilateral bending, rotation, or undulation of the flagellum (Nultsch, 1974).

PLASTIDS AND PYRENOIDS

The plastid (chloroplast) is the site of photosynthesis in eukaryote cells. Photosynthetic function is considered in Chapter 2. The physical components of the system are considered here. The basic structure of the apparatus is a system of flattened membranous vesicles and a surrounding matrix (stroma). The flattened vesicles are sac-like and have a single membrane. They are called thylakoids (Bisalputra, 1974) and are the site of the photosynthetic light reactions. Carbon dioxide fixation occurs in the chloroplast matrix. The matrix and thylakoids are enclosed by a pair of limiting membranes to give the chloroplast organelle.

Besides the thylakoids, the following structures may be enclosed by the limiting membranes: ribosomes, DNA molecules (genophore), plastoglobuli, eyespots, pyrenoid, and crystalline inclusions. The arrangement of some of the components is shown in Figure 6 and the features of seaweed chloroplasts are shown in Table 2.

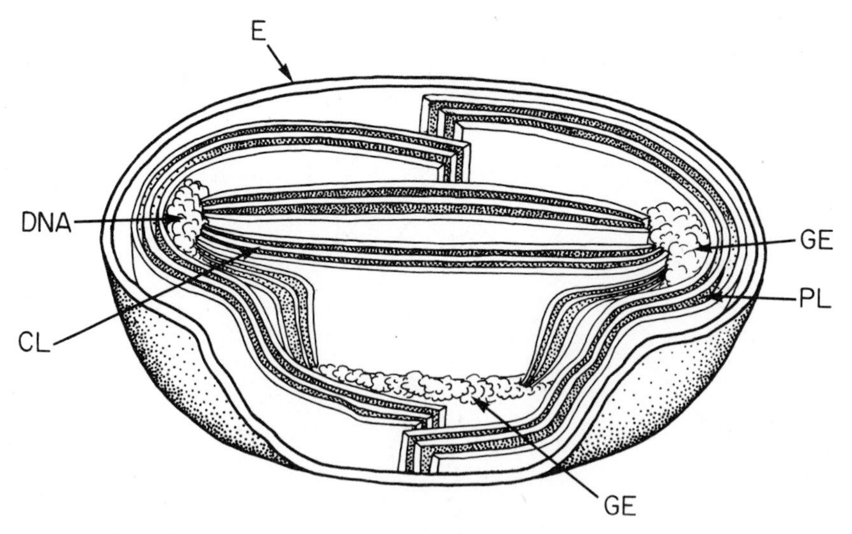

Figure 6. Three-dimensional reconstruction of a chloroplast section showing arrangement of central thylakoid lamellae (CL), peripheral thylakoid lamellae (PL), and genophore (GE) with DNA, enclosed within the chloroplast envelope (E). After Bisalputra (1974). By permission of Blackwell Scientific Publications, Ltd.

Table 2. Plastid (chloroplast) characteristics in Chlorophyta, Phaeophyta, and Rhodophyta

Division	Number of thylakoids per lamella	Chloroplast lamellae				Envelope		Reserve Products	
		Girdle lamella present	Interconnections between lamellae	Grana formed	Main chlorophylls	Number of membranes around chloroplast	Endoplastic reticulum connections	Main food reserve	Starch within chloroplast
Chlorophyta	2 many fused	–	+	+	*a, b*	2	–	Starch	+
Phaeophyta	3	+	+	–	*a, c*	4	+	Mannitol Laminarin	–
Rhodophyta	1	–	–	–	*a*	2	–	Starch	–

After Dodge (1973), with permission, from *The Fine Structure of Algal Cells*. Copyright © 1973 by Academic Press Inc. (London) Ltd.

Considerable phylogenetic significance has been placed on the stacking of thylakoids. In the red algae thylakoids occur singly in the stroma, whereas in the brown algae they are grouped (but not fused) in stacks of three. Finally, in the chlorophytes, there are stacks of two to six many-fused thylakoids with inter-stack lamellar connections. These are similar to those found in higher plants and are called grana.

In the red algae the characteristic photosynthetic phycobilin pigments are present in phycobilisomes. These are particles 35 nm across and located on the outer faces of the thylakoid membrane.

A pyrenoid is a differentiated region of a chloroplast that is mainly occupied by a proteinaceous matrix (Dodge, 1973). The pyrenoid appears to function in the conversion and translocation of early photosynthates in chloroplasts. The pyrenoid may be embedded in the chloroplast or occur as a projection from the chloroplast. The embedded pyrenoid of green algae is surrounded by a sheath of starch. In the brown algae pyrenoids are found in morphologically simple members or in the early developmental stages of complex forms. In these algae the pyrenoid projects from the chloroplast on a short stalk and a common membrane encloses the two. Around the pyrenoid and outside this first membrane system is a cap of carbohydrate that is further enclosed by another limiting membrane. The red algal pyrenoid is embedded and single thylakoid lamellae enter it.

Plastoglobuli (Lichtenthaler, 1968) are lipid granules scattered in the chloroplast. These granules appear to act as a pool of lipid reserve for membrane synthesis in chloroplast.

This completes the account of chloroplast structure. Pigmentation and photosynthetic functions are treated in the next chapter. The next organelle to be considered, the eyespot, occurs as part of the chloroplast in the motile stages of green and brown algae. It is absent from the Rhodophyta.

EYESPOTS

An eyespot is an organelle consisting of a number of osmiophilic globules containing carotenoid pigment. In motile green algal cells the eyespot is part of the chloroplast, and is about one-half cell's length distant from the insertion of the flagella (Figure 7). In the Phaeophyta the eyespot has a much closer association with the flagellum, although it is still part of the chloroplast (Figure 8). The flagellum has a granular rounded swelling that fits into the concave outer face of the eyespot.

Although we do not know how flagella work, we know what they do. With eyespots, we do not even know what they do. It has been

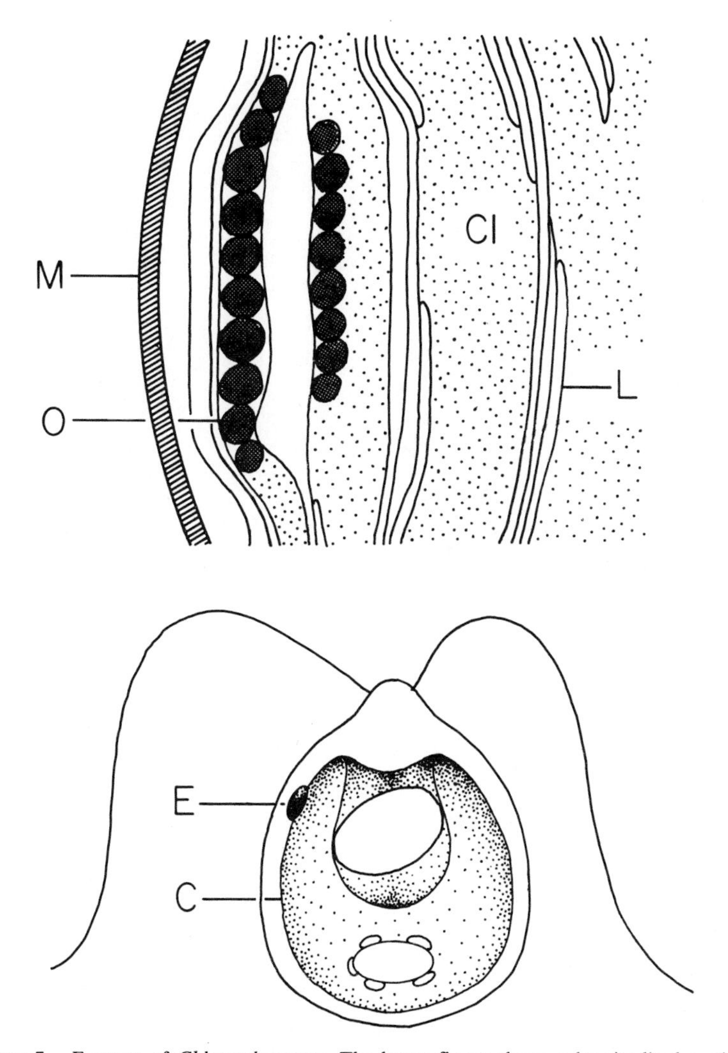

Figure 7. Eyespot of *Chlamydomonas*. The lower figure shows a longitudinal section of whole cell showing the eyespot (E) embedded in the chloroplast (C). The upper figure shows an enlarged portion of the sectioned chloroplast in the region of the eyespot. M, membrane; O, osmiophilic granule; Cl, chloroplast stroma; L, chloroplast lamellae. After Dodge (1973), with permission, from *The Fine Structure of Algal Cells*. Copyright © 1973 by Academic Press Inc. (London) Ltd.

suggested that they shade the flagellar swelling of brown algae and that the swelling is a photoreceptor. In this way the direction of illumination may be discerned. Alternatively the swelling may shade the eyespot. The function of the eyespot of green algae is difficult to deduce since it has no obvious connection with the flagellum.

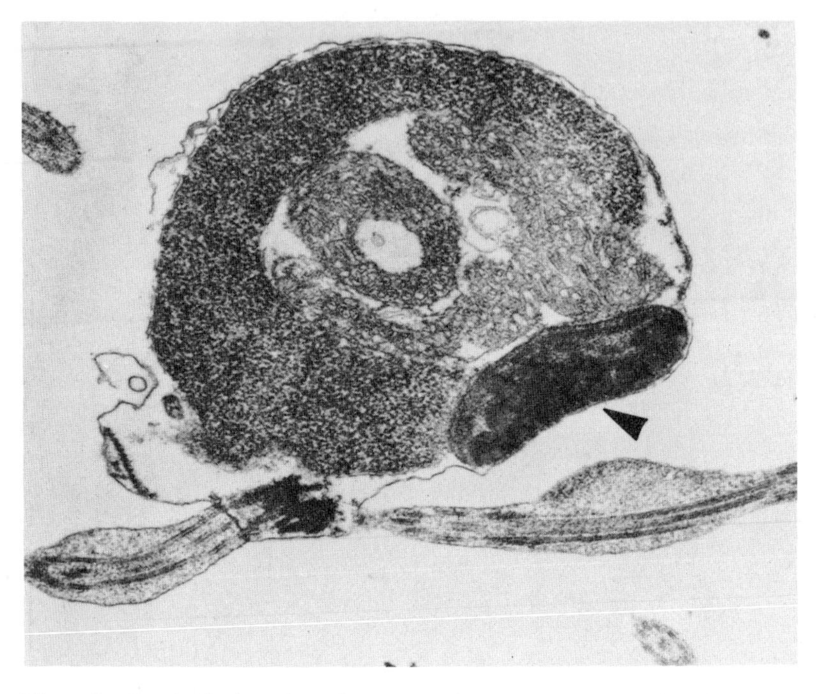

Figure 8. Longitudinal section of a sperm of *Fucus serratus* showing the eyespot (arrowed) and adjacent posterior flagellum. × 23,000. From Dodge (1973), with permission, from *The Fine Structure of Algal Cells*. Copyright © 1973 by Academic Press Inc. (London) Ltd.

NUCLEI

In seaweeds the interphase nucleus is typically eucaryotic. There is a double membrane surrounding the structure. The membrane is continuous with the cellular endoplasmic reticulum (ER) and is perforated by numerous pores. According to Bouck (1969), the gap between the two membranes of the nucleus, the perinuclear space, has an important role in flagellar hair formation. In the brown algae the nuclear membrane connects with the ER surrounding the chloroplast.

Inside the nuclear membrane are one or two RNA-rich nucleoli and chromatin, which fills the major part of the nucleus.

Very few fine structural studies of nuclear division in algae have been undertaken and most of these have concerned unicellular forms. Basically the process seems to be similar to that of most higher plants. In *Ulva mutabilis* (Løvlie and Bråten, 1970) for example, centrioles, resembling flagellar basal bodies in structure, appear to organize the formation of spindle microtubules. The chromosomes attach to the microtubules by means of centromeres and go through the well-known movements of anaphase. In *Laminaria saccharina* (J. Penny, personal communication) the centrioles appear to arise de novo. The red algae lack centrioles as well as flagella. In *Membranoptera alata*, McDonald (1972) demonstrated a structure superficially resembling a centriole, but without the typical cartwheel orientation of microtubules. This structure was termed a *polar ring*. A polar ring appears at each pole during early mitosis, and in some cases microtubules appear to be focused on the polar ring. The fine structure of meiosis in the very few seaweeds that have been examined appears similar to that of higher plants (Bråten and Nordby, 1973; Kugrens and West, 1972).

Cytokinesis is the division of the cell cytoplasm and this may or may not occur at the same time as nuclear division. The fine structure of cytokinesis of green algae has been extensively studied and reviewed by Pickett-Heaps (1975) and Stewart and Mattox (1975). Much consideration has been given to the phylogenetic and taxonomic significance of the findings. Phylogeny and taxonomic aspects of seaweeds are not under consideration here and the reader is referred to the literature listed above for details.

OTHER CELL COMPONENTS

The transport system consists of the perinuclear space, the lumen of the ER, and the Golgi or dictyosome apparatus. The Golgi and ER of the zoospore of *Enteromorpha* are shown in Figure 9. The Golgi functions as part of the internal transport system of the cell and also in the formation and packaging of substances for extracellular transport (Evans, 1974). There is strong evidence of involvement of the Golgi in polysaccharide synthesis in brown algae. Sulfated polysaccharides, such as fucoidan, appear to be synthesized in these structures, as is shown by the sophisticated EM-autoradiography studies of Evans, Simpson, and Callow (1973). In addition, the adhesive required for attachment of algal spores appears to be synthesized in the Golgi (Baker and Evans, 1973; Chamberlain and Evans, 1973). The Golgi is also involved in wall formation, so that clearly this is a versatile organelle.

Other organelles found in seaweeds are mitochondria and microbodies. The structure and formation of mitochrondria are well known and are not considered further.

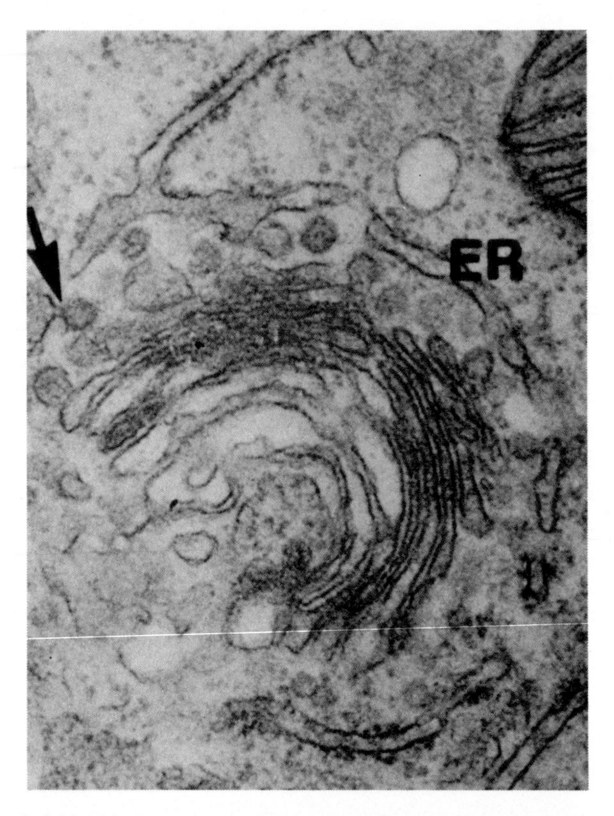

Figure 9. Golgi body of a zoospore of *Enteromorpha* partially surrounded by an arc of endoplasmic reticulum (ER). The arrow points to the small vesicles which are abstricted from the ER and are believed to transport material to the Golgi. × 75,000. From Evans (1974). By permission of Blackwell Scientific Publications Ltd.

Microbodies occur in brown algae (Bouck 1965) and are common in higher plants. The organelles are bound by a single membrane and are granular in content. They are thought to be associated with photorespiration because they contain glycolate oxidase and catalase, which are both associated with photorespiratory activity.

For detailed reviews of algal fine structure the reader is referred to Bisalputra (1974), Dodge (1973), Evans (1974), and Pickett-Heaps (1975).

Chapter 2
Cell Function

Cell function is taken here to mean cell physiology, which according to Giese (1973) encompasses the life activities of nutrition, response to environment, growth, cell division, and differentiation. This is clearly a vast field, and one chapter of a small book cannot adequately review it. Many aspects of these life activities are common to all organisms. Since this is a book about seaweeds, the field can be narrowed by considering unique features of these plants or those aspects of cell physiology that they best exemplify.

Nutrition, in the broadest sense, is an interesting aspect of algal biology. These organisms are second only to bacteria in their nutritional diversity. In this chapter emphasis is placed on the range of nutritional processes. Photosynthesis and its associated pigment systems are given separate treatment. Only two aspects of the response of cells to their environment are considered here, the response of photosynthetic pigments to light and ionic relations with the environment. Treatment of cell growth is restricted to the growth of cell walls and their morphogenetic importance. Other aspects of development, such as establishment of polarity and growth hormones, are included in Chapter 4. The physiology of cell division is not included, and consideration of cellular differentiation is restricted to one remarkable organism, *Acetabularia*.

NUTRITION

The nutrition of such physiologically diverse organisms as algae can be confusing. One way of reducing this confusion is to consider energy sources and hydrogen donors separately, as shown in Table 3. Most green plants exhibit photolithotrophy, in which the energy source is light and the photosynthetic hydrogen donor is water. The function of the hydrogen donor is considered later in the section, "Photosynthe-

Table 3. Energy sources and hydrogen donors in Chlorophyta

Hydrogen donor	Energy sources: Chemical (*Chemotrophs*)	Energy sources: Light (*Phototrophs*)
Inorganic (autotrophs)	Chemolithotrophy. Rare, eg., *Chlamydomonas moewusii.*	Photolithotrophy. H_2O or H_2 used as H donors. Common, eg., *Ulva* sp.
Organic (heterotrophs)	Chemo-organotrophy. Facultative form is common. Sugars and acetic acid common organic sources.	Photo-organotrophy. Probably common, eg., *Desmarestia* and *Bonnemaisonia*. Use sugars and acetate in light.

sis." Some algae (e.g., *Ulva* and *Ascophyllum*) have the ability to utilize elementary hydrogen in place of water as a hydrogen donor. The overall photosynethtic changes for these two photolithotrophic reactions are:

$$CO_2 + 2H_2O \rightarrow (CH_2O) + H_2O + O_2$$
$$CO_2 + 2H_2 \rightarrow (CH_2O) + H_2O$$

The second process, unlike the first, takes place under dark anaerobic conditions and depends on the presence of a latent hydrogenase system.

Autotrophic algae that use chemical oxidation of simple inorganic compounds as an energy source are rare. To date the process has been found in only a few unicellular green algae. Chemolithotrophy is of course common in bacteria and forms the basis of biogeochemical cycles.

The use of organic hydrogen donors (heterotrophy) is common among the algae as a whole. There are two types of heterotrophy, chemo-organotrophy (chemical energy source) and photo-organotrophy (light energy source). Most of the algae that have been examined for heterotrophic capability are microscopic (Droop, 1974) and there is little information about seaweeds in this regard. It seems likely that photo-organotrophy is common. Nakahara and Tatewaki (1971) found that in the large brown alga *Desmarestia liqulata* glucose was required for normal morphological development. *Trailliella* has the ability to use acetate in the light.

At present true chemo-organotrophy has not been reported for seaweeds, and until it is the process cannot be used to explain seaweed growth in areas of feeble illumination.

Seaweeds have nutritional requirements beyond hydrogen donors and a source of energy. Like higher plants, there is a requirement for a wide variety of inorganic ions. Unlike most higher plants, most algae require vitamins for normal growth. This vitamin requirement is called auxotrophy (Provasoli and Carlucci, 1974). Vitamin B_{12} is a growth requirement for red, green, and brown algae. Thiamine is required by some green algae.

The following elements are considered to be macronutrients for algae: sulfur, potassium, calcium, magnesium, carbon, nitrogen, and phosphorus (O'Kelley, 1974). Sulfur is required for S-bonds in proteins and for the production of sulfated polysaccharides found in seaweeds. Potassium has a general role as an enzyme activator, and magnesium is required in the construction of the chlorophyll molecule. Calcium ions play a part in the maintenance of cell membranes and in the construction of cell walls. Carbon is obviously required for carbohydrate assembly, and in aquatic systems is present as CO_2 in physical solution and as the ions of bicarbonate and carbonate. Organic carbon may also be utilized, although it is not present in sufficient quantity in natural sea waters to be a major source of supply. Nitrogen is of course an important constituent of cellular proteins and an absolute requirement for plant growth.

Phosphorus is required in the energy generation and transfer processes of the cell. Other elements required in small quantities are termed micronutrients. They occur in the third and fourth group of the periodic table. Of these, all algae seem to require iron, manganese, copper, zinc, molybdenum, and chlorine. Iron is a constituent of cytochrome and thus of vital importance in the cellular electron transfer system. Manganese has an important role in the O_2-evolving system of photosynthesis (photosystem II, as defined in the next section, "Photosynthesis"), and copper is probably essential for photosystem I (defined in "Photosynthesis") activity. Zinc plays a part in RNA formation in preserving the structure of ribosomal components. Molybdenum seems to be required in nitrogen uptake processes, and chlorine (as chloride) is essential for the light reaction of photosynthesis and for ATP formation.

Other elements required by some seaweeds are cobalt (as Vitamin B_{12}), iodine, and boron. Iodine is required by several red and brown algae. In kelps 1% of the dry weight may be iodine. Its metabolic role is not clear and neither is that of boron.

Ion Uptake

It must be realized that the previously listed elements are usually taken up in an ionic form. The elementary states can be lethal, e.g., chlorine and iodine. Within the algae there are many species that possess unusually large cells. For instance in *Nitella*, a green filamentous form, the cells may be up to 15 mm long and several millimeters in diameter. This has provided excellent experimental material for studying the uptake of ions by cells. In the green seaweed *Valonia* the concentration of potassium is over 40 times as high as that in the surrounding water (Davson, 1970). There must be some form of active transport to maintain such a steep gradient. The potassium is in an ionic form. This is shown by measurements of cell conductivity. If a cell is deprived of O_2 (presumably shutting down active transport), the K^+ leaks out of the cells. The site of active transport is the cell membrane. It is also known that sodium ions are pumped out of cells. This creates a favorable cation gradient for the

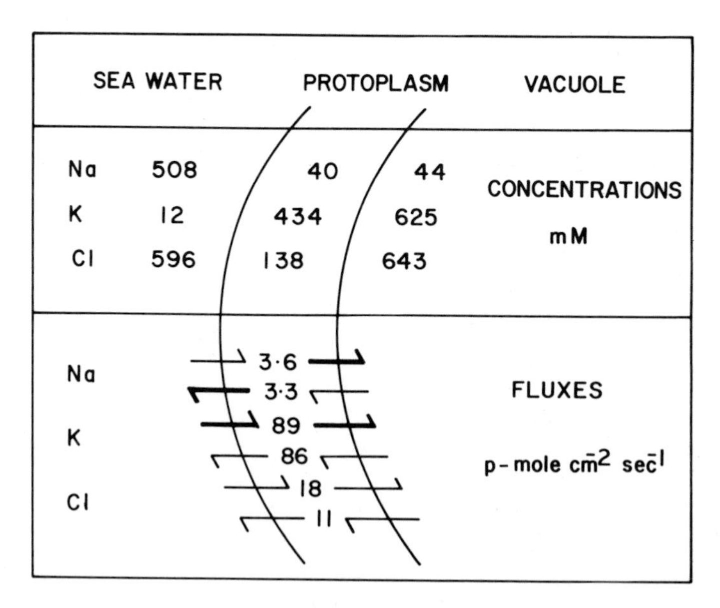

Figure 10. Ionic relations of the siphonous green alga *Valonia*. Provisional scheme showing ionic concentrations and fluxes. Broad arrows indicate active transport. After Gutknecht and Dainty (1968).

passive uptake of potassium (which is nevertheless accumulated by active transport).

Chloride appears to be taken up passively into the cell of *Valonia* (Gutknecht and Dainty, 1968). The ionic relations for Na^+, K^+, and Cl^-in *Valonia* are shown in Figure 10. Nitrate and iodide seem to be taken up actively in *Valonia*. In the freshwater algae *Nitella* and *Chara* active influx of PO_4^- and HCO_3^- at the plasmalemma is suggested. The active pumping of charged ions creates an electrochemical potential across the plasmalemma. In *Bryopsis* this is -70 mV. The mechanism of ion transport is unknown. Most theories postulate a carrier molecule that ferries ions across the membrane.

PHOTOSYNTHESIS

Phototrophic algae are photosynthetic. In photosynthesis light energy is converted into potential chemical bond energy of ATP. CO_2 is reduced to carbohydrate. Energy for reduction is provided by light energy, and hydrogen is provided by the photolysis of water. The process is complicated and research is of the "frontiers of science" variety. However we can summarize the process briefly, as in Figure 11. The process of photosynthesis is considered here in two parts. The first deals with the photochemical aspect, which in the scheme of Figure 11 is all of the processes up to NADPH formation. The second is CO_2 fixation, shown at the top of the scheme as $CO_2 \rightarrow (CH_2O)$.

The light quanta striking the photosystems must be absorbed to carry out the photochemical process. This absorption is carried out by a wide variety of photosynthetic pigments. The types of pigments found in seaweeds and some of their characteristics are shown in Table 4. The chemical characteristics of the pigments are not of concern here and are reviewed in detail by Govindjee and Braun (1974). However, the physical characteristics, in terms of their selective absorption of light of varying wavelengths, are relevant and some of the details are given in Table 4.

The first question for consideration is: how does a pigment molecule in vivo react when it absorbs a quantum of light? It appears that an electron is lifted out, leaving the pigment molecule positively charged. The lifted electrons are moved from one carrier to another in the process of electron transport. Changes in potential energy in the transfer are coupled with the formation of ATP. It is important to realize that energy can be transferred between pigment molecules and

Table 4. Photosynthetic pigments in algae

A. The Chlorophylls

Type of chlorophyll	Characteristic absorption peaks		Occurrence
	In organic solvents (nm)	In cells (nm)	
a	420, 662	435, 670–680 (several forms)	All algae
b	455, 644	480, 650 (two forms?)	Chlorophyta
c	444, 626	Red band at 645	Diatoms and Phaeophyta
d	450, 690	Red band at 740	Reported in some Rhodophyta (?)

B. The Carotenoids

I. Carotenes

Types of carotenoids	Characteristic absorption peaks (nm)	Occurrence
α-carotene	In hexane, at 420, 440, 470	In Rhodophyta and in siphonaceous Chlorophyta it is the major carotene
β-carotene	In hexane, at 425, 450, 480 (the 480-nm band may be shifted to 500 nm in vivo	Main carotene of all other algae

II. The Xanthophylls

Types of Xanthophyll	Characteristic absorption peaks (nm)	Occurrence
Lutein	In ethanol, at 425, 445, 475	Major carotenoid of Chlorophyta and Rhodophyta
Fucoxanthin	In hexane, at 425, 450, 475 (in vivo, absorption extends to 580 nm)	Major carotenoid of diatoms and Phaeophyta

C. The Phycobilins

Types of Phycobilin	Absorption peaks (nm)	Occurrence
Phycoerythrins	In water and in vivo, at 490, 546, 576	Main phycobilin in Rhodophyta also found in some blue-green algae
Phycocyanins	In water and in vivo, at 618	Main phycobilin of blue-green algae, also found in Rhodophyta
Allophycocyanin	In phosphate buffer (pH 6.5) and in vivo, at 654	Found in blue-green algae and Rhodophyta

After Govindjee and Braun (1974). By permission of Blackwell Scientific Publications Ltd.

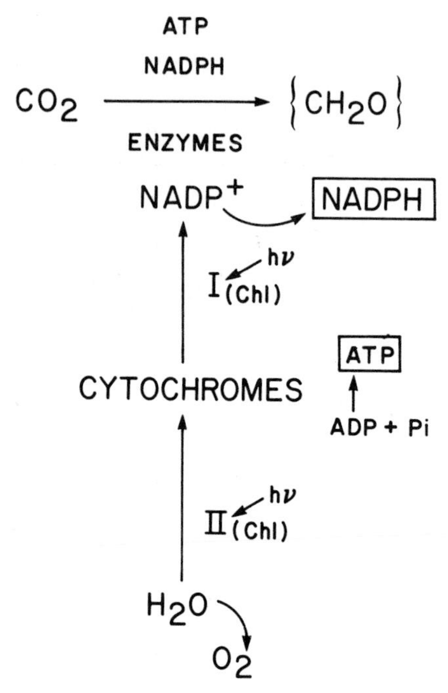

Figure 11. Photosynthetic transfer of electrons from water to carbon dioxide involving two photosystems (I and II). ATP, adenosine triphosphate; $NADP^+$, nicotinamide adenine dinucleotide phosphate; NADPH, reduced $NADP^+$; Chl, chlorophyll; (CH_2O), carbohydrate moiety; hν, light quanta. From Govindjee and Braun (1974). By permission of Blackwell Scientific Publications Ltd.

that the pigments may be similar or dissimilar. This brings us to the role of accessory pigments.

All photosynthetic plants have chlorophyll *a*, which is said to be the primary photosynthetic pigment. In many algae the green chlorophyll is masked by other pigments—brown fucoxanthin in brown algae and red phycobilin in red algae. If the various pigments are extracted in organic solvents, they show differential absorption of various wavelengths of visible light (Table 4). This is expected since pigments tend to absorb colors complementary to their own. The absorption capabilities of a pigment are expressed in the absorption spectrum. The action spectrum of photosynthesis expresses relative rates of photosynthetic O_2 liberation across the light spectrum. That the carotenoids and phycobilins participate in photosynthesis was first shown by the concordance of the photosynthetic action spectrum of red and brown algae and their absorption spectra for mixtures of extracted pigments. It is now known that chlorophyll *a* carries out primary photochemical conversion and that, when other pigments absorb light quanta, energy is transferred to chlorophyll *a*.

It now appears that the pigment molecules of algae are arranged in two systems, known as photosystems I and II, shown in Figure 12 for green algae. In the brown algae chlorophyll *c* replaces chlorophyll *b*, whereas in the red algae phycobilins replace it. Chlorophyll *a*, shown as chlorophyll *a* 670, 680, 685, 705 in Figure 12, exists in several spectroscopically distinguishable forms. $P_{680\text{-}690}$ and P_{700} are the chlorophyll pigment traps. Z and X have not been identified. In green algae photosystem I is situated on the outer side of the thylakoid membrane and photosystem II on the inner side. In red algae the

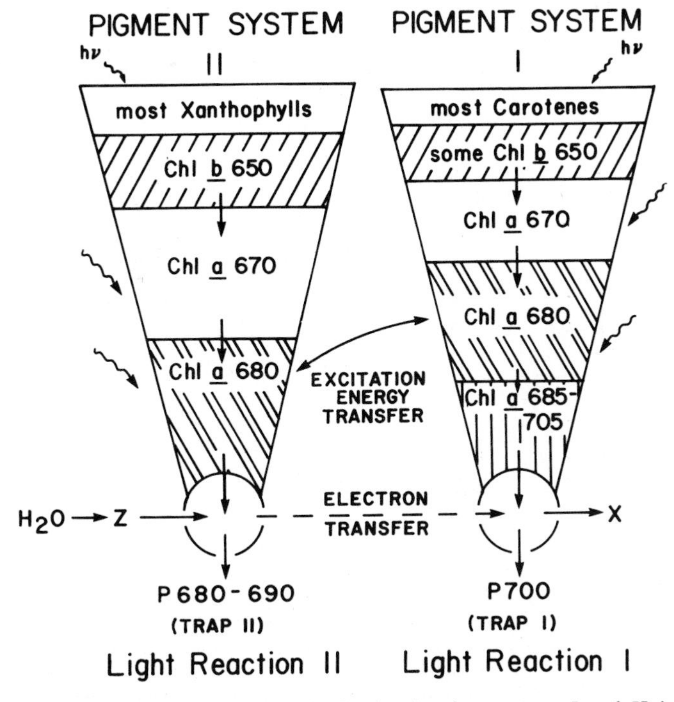

Figure 12. Hypothesis for pigment distribution in photosystems I and II in green algae. Chl *b*, chlorophyll *b*; Chl *a*, chlorophyll *a*; numbers after Chl *a* indicate absorption maxima at the red end of the light spectrum; Z, primary electron donor of system II; X, primary electron acceptor of system I; P, pigment trap. Modified after Govindjee and Braun (1974). By permission of Blackwell Scientific Publications Ltd.

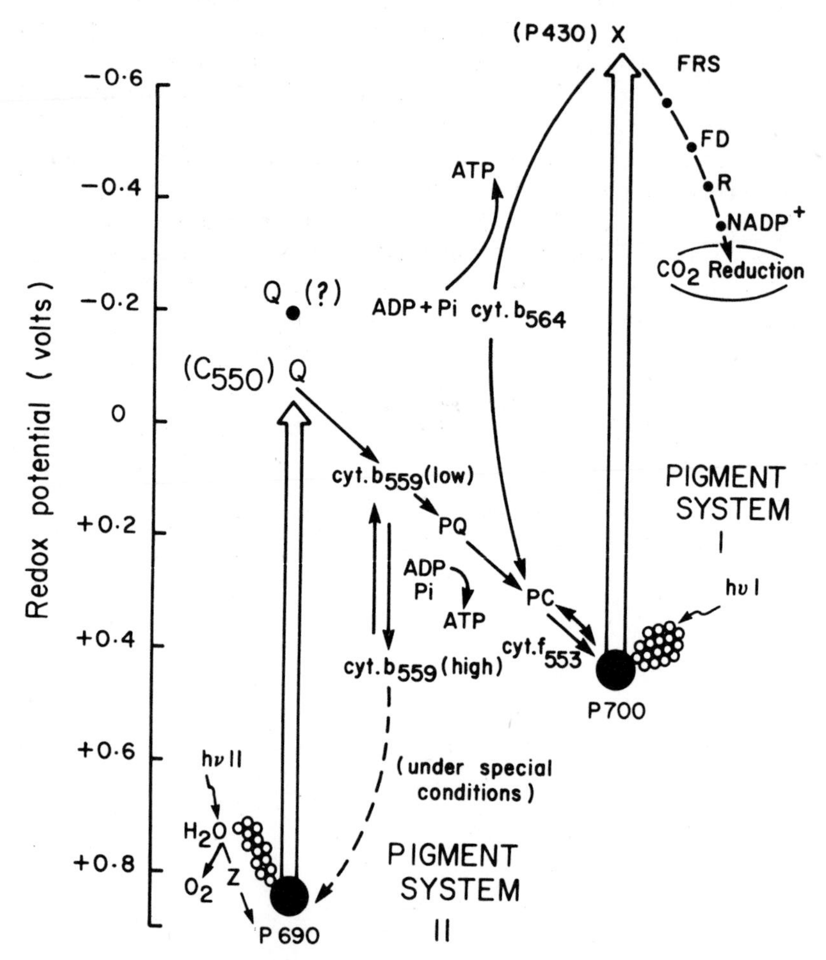

Figure 13. Scheme of photosynthesis showing electron flow. The two double arrows from P_{690} to Q and P_{700} to X represent light reactions; all other arrows represent dark reactions. Electrons flow noncyclically from H_2O (lower left) to $NADP^+$ (upper right) and cyclically in Pigment System I from X to system intermediates. A similar cyclical flow in PS II, shown as a dashed line, may also operate. Q (C_{550}), primary electron acceptor for PS II; PQ, plastoquinone; cyt b_{559} and cyt b_{564}, cytochromes *b*; cyt f_{553}, cytochrome *f*, PC, plastocyanin; FD, ferredoxin; R, FD-$NADP^+$ reductase. Other symbols as in Figure 12. After Govindjee and Braun (1974). By permission of Blackwell Scientific Publications Ltd.

phycobilins are located in bodies known as phycobilisomes, which are attached to the thylakoid membrane. The physical arrangement of pigment systems in red algae is not fully understood.

The excitation energy transfers through the pigment systems are light reactions. Electron transfer through $H_2O \rightarrow$ Z (primary electron donor) $\rightarrow P_{680\text{-}690} \rightarrow P_{700} \rightarrow$ X is termed the dark reaction. The dark reaction involves a series of steps that are shown in an expanded form in Figure 13.

One of the end products of photochemical electron transport is reducing power in the form of $NADPH_2$, which is used for CO_2 fixation. Figure 14 shows the photosynthetic carbon reduction cycle. A C-4 acid cycle has not been identified in algae as yet (Raven, 1974). The Calvin cycle is a source for carbohydrate formation and for the carbon components of lipids, amino acids, pyrimidines, and porphyrins.

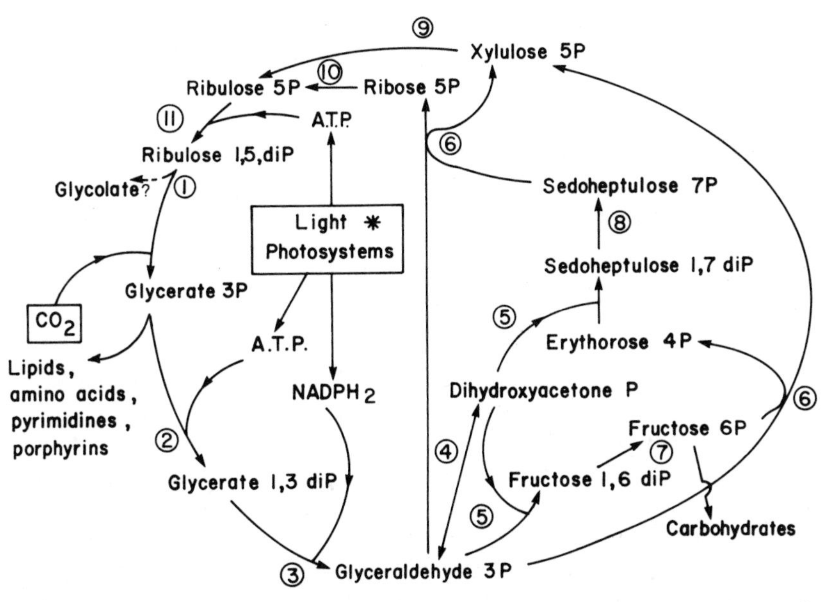

Figure 14. Photosynthetic carbon reduction cycle. *, catalytic role of CO_2. Enzymes of the cycle are referred to by number: 1, ribulose diphosphate carboxylase; 2, 3-phosphoglycerate kinase; 3, phosphoglyceraldehyde dehydrogenase; 4, phosphotriose isomerase; 5, fructose diphosphate aldolase; 6, transketolase; 7, fructose-1,6-diphosphate-1-phosphotase; 8, sedoheptulose-1,7-diphosphate-1-phosphotase; 9, phosphopentose epimerase; 10, phosphoribose isomerase; 11, phosphoribulokinase. After Raven (1974). By permission of Blackwell Scientific Publications Ltd.

This completes the present treatment of nutrition and photosynthesis in algal cells. Considerations of the response of cells to environment have been limited here to the response of cells to light and the exchange of ions across the cell membrane. This brings us to the last of the life activities of cells, cell growth and differentiation.

CELL GROWTH AND DIFFERENTIATION

The growth of plant cells appears to result from a yield of the wall to an internal turgor pressure (difference between the osmotic pressure of intra- and extracellular fluids) (Green, 1969). In the green alga *Chaetomorpha* the turgor pressure is 13–14 atmospheres (Gutknecht and Dainty, 1968). In most marine algae the high osmotic pressure is due to the active uptake of Cl^- and other anions. In some forms, such as *Valonia* and *Chaetomorpha*, the high internal salt content results from the inward pumping of cations, principally K^+.

The amount of enlargement resulting from yield to turgor pressure can be enormous. In *Antithamnion plumula* the increase in cell volume during growth is 48,000-fold (Dixon, 1971). Many filamentous algae have long cylindrical cells. The stress pattern caused by internal turgor pressure in elongating cylindrical cells favors increase in girth (Green, 1969). How is it then that the cells remain long and narrow? In certain algae the answer appears to lie in the arrangement of cell wall microfibrils. In *Nitella* there is an overall hoop-like arrangement of microfibrils that strongly reinforces the wall so that most expansion occurs in the axial direction. As a cell elongates the wall tends to thin. Increase in thickness is brought about by deposition of microfibrils on the internal surface, and these are predominantly in a transverse orientation. After deposition a group of microfibrils undergoes thinning and passive realignment in the axial direction.

If, in cells with a multi-net configuration, the orientation of microfibrils determines the pattern of cell growth, what controls the ordered deposition of microfibrils? It is Green's contention that microtubules play a key role here (Green, 1969; Green, Erickson, and Richmond, 1970). Microtubules are mentioned in Chapter 1 in connection with flagellar construction and the formation of mitotic spindles. These proteinaceous tubules also form a cytoplasmic "skelton." In *Nitella* the orientation of microtubules in the cytoplasm below the cell wall appears to follow the orientation of microfibrils. Fortunately there is experimental evidence to test this correlation. The drug colchicine interferes with the orientation of microtubules in cells. Thus it will disrupt cell division by preventing spindle formation. If the drug is applied to small, growing cells of *Nitella* it will produce a random orientation of cytoplasmic microtubules, which in turn appears to result in a random deposition of cellulose microfibrils. This has an interesting effect on cell shape—the young cells assume a spherical appearance. This experiment has two important conclusions: first, the orientation of microfibrils controls cell shape, and second, the orientation of microtubules controls the pattern of microfibril deposition. Unfortunately, this mechanistic picture of cell growth seems to be of limited application to the algae. Green et al. (1970) take their theories much further. If microtubule alignment controls microfibril pattern, what controls the orientation of the tubules? They propose three possible orienting forces: crystallization, electrophoresis, and strain alignment:

a. Crystallization: There is a tendency for microtubules to align with organization centers such as centrioles. Green conceives of a hierarchy of crystallizations, which would account for microtubule-microfibril orientation by a type of template inheritance.
b. Electrophoresis: It is well known that in growing tip cells there is a small electrical current flowing from one end to another. This current may be involved with the modification of tip growth zones responding to light gradients.
c. Strain alignment: In *Nitella* there is a transversely directed stretch in the base of the apical cell, and this is correlated with the subsequent appearance of a transverse wall structure. It may be that random microtubules receive their initial organization through this transverse strain.

As mentioned earlier, it is not always possible to explain the direction of cell growth in terms of microfibril orientation. In fact, most filamentous red and brown algae have a random orientation of fibrils. In *Chaetomorpha*, a green alga with cylindrical cells, the fibrils are in the "crossed-fibrillar" orientation. It is not possible in these cases to propose a mechanistic explanation of microfibrillar control of cell expansion.

In some algae it seems that microtubules are not responsible for microfibril alignment. Work on *Cladophora* and *Chaetomorpha* (Robinson and Preston, 1971; Robinson, White, and Preston, 1972) has shown that zoospores lack cell walls. After a spore settles, arrays of granules, probably enzyme complexes, form on the outer face of the plasmalemma. Cellulose synthesis takes place in the granules. Mackie and Preston (1974) reject the theory of fibril synthesis by the cytoplasmic microtubular system. Green et al. (1970) on the other hand feel that fibril orientation "cannot be properly viewed entirely in terms of particles." This point has yet to be resolved.

Cell differentiation is poorly understood in algae as a whole, but has been extensively studied in the remarkable seaweed *Acetabularia* (Figure 15). This plant consists of a single uninucleate cell which may be 5–6 cm long. The cell is differentiated into three parts: an apical cap, a stalk, and attaching rhizoids. The interaction between nucleus and cytoplasm in controlling the differentiation of these three cellular components was first studied in the 1930s. Elegant surgical procedures by Hämmerling and his co-workers continued for many years and were reviewed in 1963 (Hämmerling, 1963).

The nucleus of *Acetabularia* is situated in the rhizoids, thus it is possible to remove the cap without affecting the nucleus. When

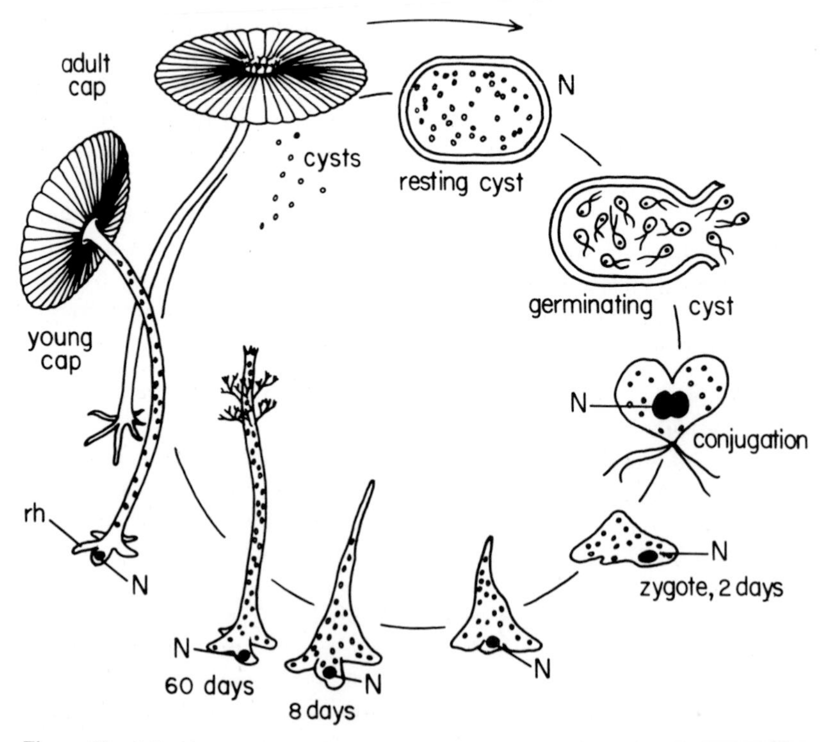

Figure 15. Life history of the green alga *Acetabularia*. rh, rhizoid; N, nucleus. After Brachet (1957). By permission, from *Biochemical Cytology*. Copyright by Academic Press Inc.

reproduction takes place the primary nucleus divides, giving 7,000–15,000 secondary nuclei. These migrate to the cap, form cysts, and are released. If the cap is removed and all of its regenerated successors, nuclear division is inhibited. Regeneration has been observed over a 3-year period with no sign of cytoplasmic or nuclear aging.

It is possible to remove the nucleus from developing cells by surgery. Surprisingly, these cells are able to form new stalks and healthy growing caps. Thus there are high morphogenetic capacities in enucleated cells. According to Hämmerling these capacities are under nuclear control and are dependent on the prior release by the nucleus of "morphogenetic substances" into the cytoplasm. These substances migrate to the cap, establishing an apico-basal concentration gradient.

This work shows that the cytoplasm controls nuclear division and that the nucleus influences morphogenetic activity of the cytoplasm. Thus nucleocytoplasmic interaction works both ways.

Acetabularia occurs as several well defined species. *A. mediterranea* and *A. crenulata* have distinctive cap morphologies. The tough nuclei of *Acetabularia* withstand transplantation between species. If a nucleus from an individual of *A. crenulata* is transplanted into the previously enucleated cytoplasm of an individual of *A. mediterranea*, a more or less pure *crenulata* cap is formed. The reverse procedure produces a *mediterranea* cap, or an intermediate. Thus the morphogenetic substances produced by the nucleus are species specific. Where intermediate caps are formed, the effect is temporary, since amputation of this cap results in the regeneration of a cap having the species-specific morphology of the implanted nucleus. Intermediate caps are formed where there is a leftover supply of morphogenetic substances from the excised nucleus. This work means that an A nucleus makes A cytoplasm from B cytoplasm. In other words the ribosomes of B can form the species-specific proteins of A.

The nature of Hämmerling's "morphogenetic substances" is still not clear. They may be long lived RNAs, and this would of course jibe with the central dogma of DNA→RNA→protein.

The foregoing is, unfortunately, the extent of our knowledge of the control of cellular differentiation in seaweeds. Some information on tissue formation exists and this dealt with in Chapter 4.

Part II
Organization in Whole Organisms

In vascular plants, cells are organized in tissues which make up the organs—leaf, stem, root. According to the earliest concepts put forward by Sachs (1882), there is a basic architectural plan common to vascular plants. In German literature this basic plan (*bauplan*) is the cormus (Sachs, 1882). All of the vast variety in the morphology of vascular plants may be reduced to homologies of the organs in the *bauplan*. This unifying concept has no equivalent in the algae. There is no common plan to the diversity of structure and no agreement on the evolutionary origins of one type from another.

Fortunately this book is concerned only with the macroscopic marine algae, among which unifying trends can be discerned. These are discussed in Chapter 3.

In Chapter 4 attention is focused first of all on the functional aspects of the growth forms of seaweeds. This discussion is only possible through the fascinating and innovative studies of Michael Neushul and his collaborators. Aspects of seaweed morphogenesis and physiological ecology are also considered. Chapter 5 deals with reproduction in seaweeds.

Chapter 3
Thallus Structure

A multicellular seaweed plant body is usually called a *thallus*. In the seaweeds the levels of morphological complexity can be resolved into five major types: filaments, heterotrichs, siphons, parenchyma, and pseudoparenchyma.

FILAMENTS

The simplest seaweed is an unbranched chain of cells. Only the attaching cell is differentiated from the rest (Figure 16). *Ulothrix flacca* (Chlorophyta) displays this type of construction. Apart from the attaching rhizoidal cell, each member of the chain has the capacity for division and the production of reproductive bodies. According to Fritsch (1935), filaments are characterized by vegetative cell division in which the parent wall of a dividing cell is retained by the daughter cells. This contrasts with the type of cell division found in unicells and colonies where the parent cell wall is ultimately cast off or gelatinized.

The simple uniseriate filament has many elaborations. Among these are the formation of apical growing points. Here cell division is restricted to the apical cell. Another modification is the formation of lateral branches. By changes in the plane of apical cell division, dichotomous branches are produced.

In all of these forms, the filaments are of one cell thickness. If, during division, new walls arise along two planes at right angles to one another, then flattened biseriate or multiseriate sheets are formed. This is shown in Figure 17 for *Percursaria* and *Ulva*. These leafy green thalli of the Chlorophyta are conspicuous on sewage-polluted shores.

SIPHONS

In the green algae some remarkable morphological forms usually known as siphons develop. These are characterized by the develop-

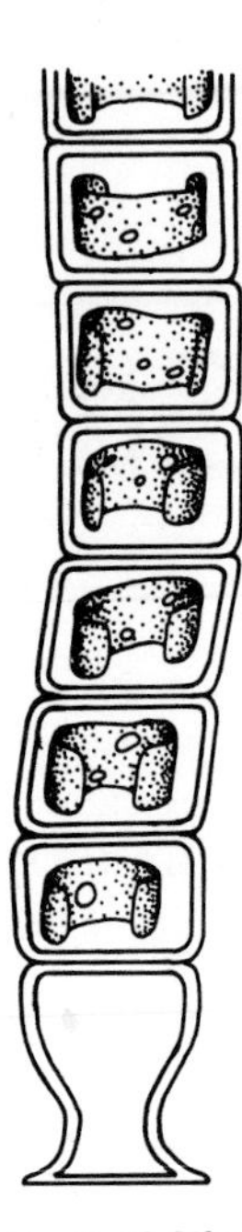

Figure 16. Simple unbranched filament of *Ulothrix* showing differentiation of basal holdfast cell. × 650.

ment of a large plant body without cross walls or septa (these may be present rarely). Elegant, aesthetically pleasing forms are common. One of these, the genus *Acetabularia*, was discussed in the last chapter. *Acetabularia* is unusual in having only a single nucleus. Most forms are multinucleate. The range of growth forms is extensive (Figure 18).

HETEROTRICHS

Heterotrichous forms are clearly related to filaments, but are distinguished from the latter by a differentiation into upright and prostrate systems (Figure 19). There may be reduction of either of these systems so that only the erect or prostrate systems are visible in the mature plant. In the latter case one may include virtually all of the encrusting seaweeds. In Figure 56 the pink zone is made up of an encrusting calcified red algae. In some of these crustose forms erect filaments are never formed except in connection with the production of reproductive

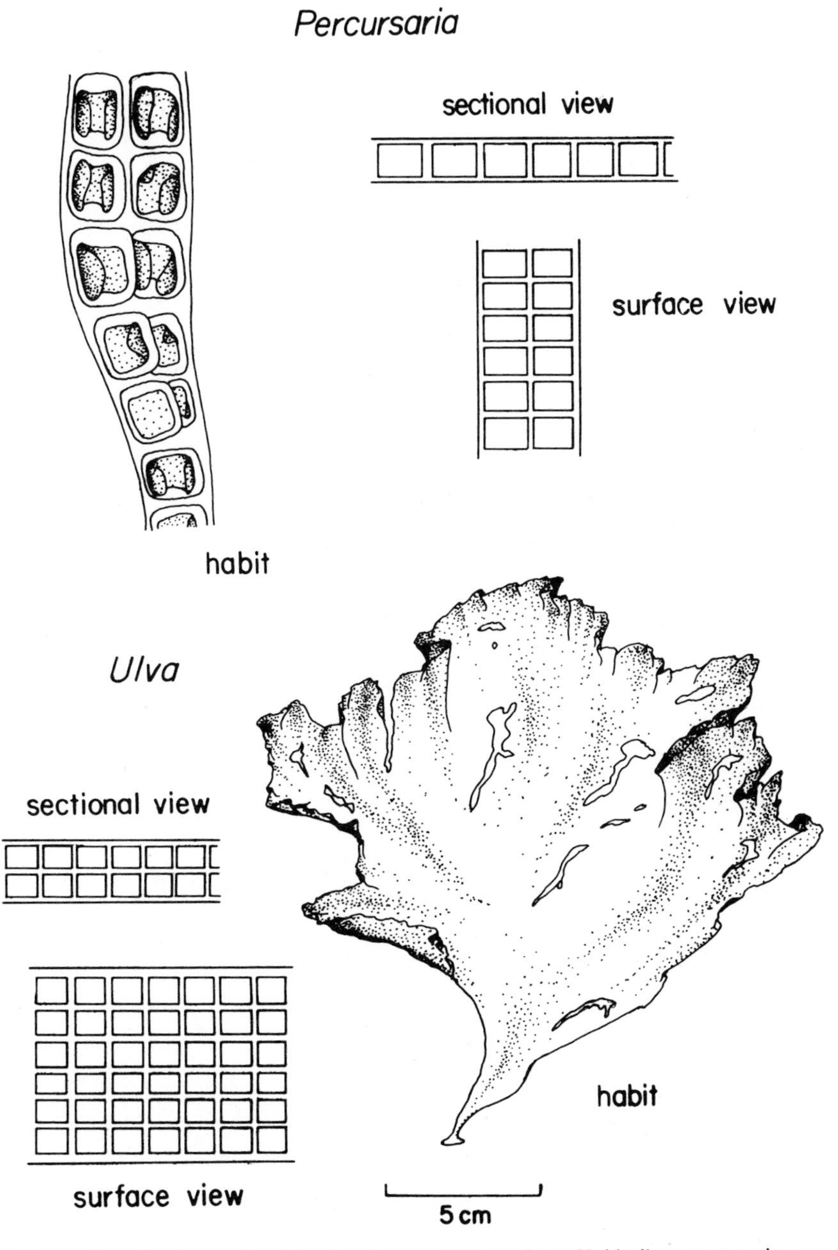

Figure 17. Biseriate and multiseriate forms of Chlorophyta. Habit diagrams are given as well as stylized sectional and surface views. *Percursaria* habit diagram × 500, *Ulva* habit diagram × 0.5. After Scagel et al. (1965), *An Evolutionary Survey of the Plant Kingdom*. Copyright © 1965 by Wadsworth Publishing Company, Inc., Belmont, Cal. By permission of the publisher.

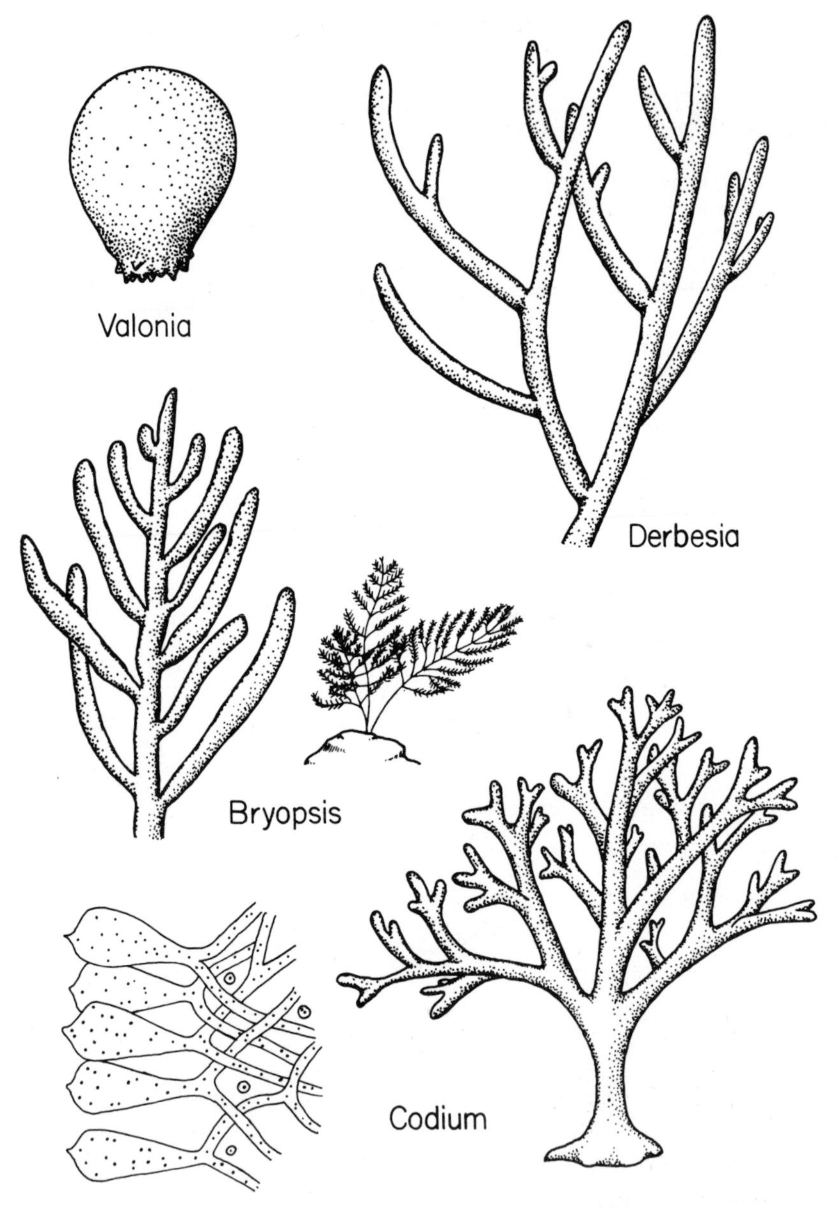

Figure 18. Siphonous forms of construction. *Valonia*, globose form, × 0.5; *Derbesia*, × 250; *Bryopsis*, pinnate form showing habit, diagram × 0.5, and detail, × 120; *Codium* pseudoparenchymatous form showing habit, × 0.25, and stylized detail of internal anatomy. After Scagel et al. (1965), *An Evolutionary Survey of the Plant Kingdom*. Copyright © 1965 by Wadsworth Publishing Company, Inc., Belmont, Cal. By permission of the publisher.

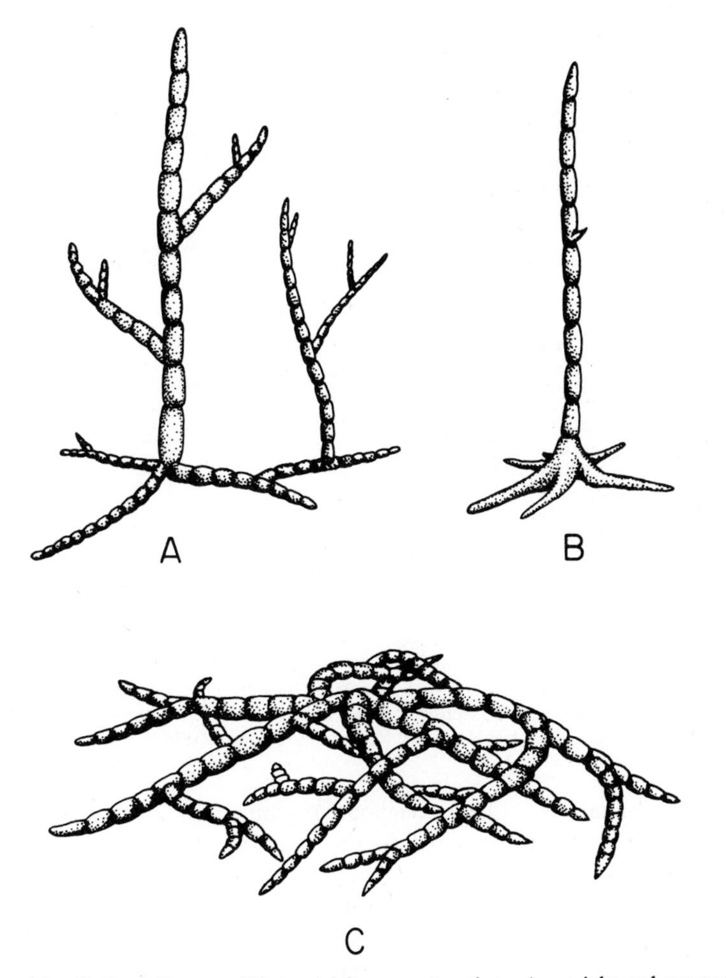

Figure 19. Stylized diagram of heterotrichous constructions. A, upright and prostrate systems well developed; B, prostrate system suppressed; C, upright system suppressed. After Gayral (1975). By permission of Doin, Editeurs.

structures. Commonly, however, there is a limited production of short upright filaments (Figure 20).

Much more common than the suppression of the erect system is the suppression of the prostrate. Among the green algae this has given rise to forms such as *Fritschiella* (Figure 21). Recently Dixon (1973) has described the occurrence of secondary heterotrichy. In *Rhizophyllis* (Rhodophyta) and some species of *Lithothamnion* the thallus

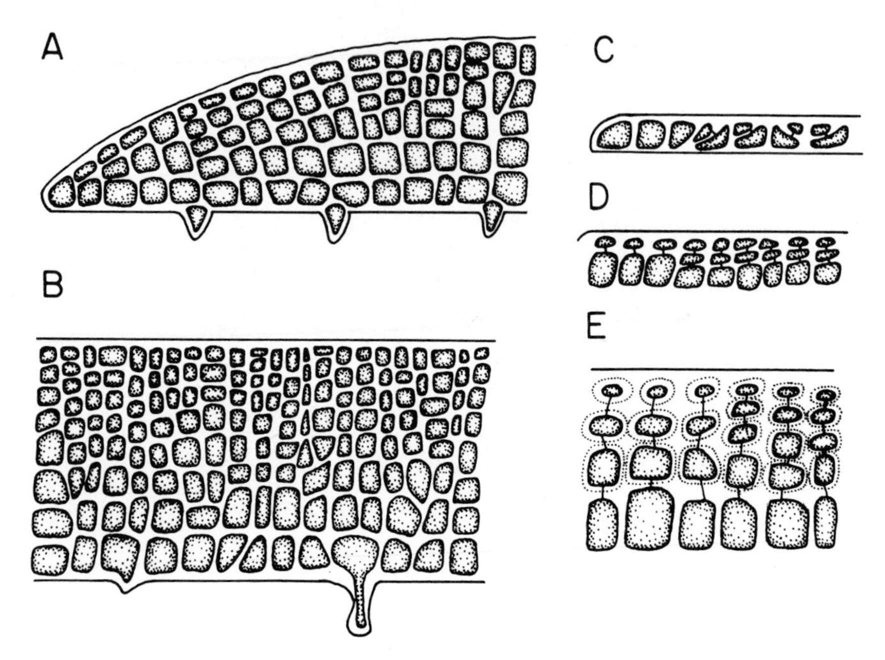

Figure 20. Crustose red algal thalli. A–B, *Peyssonelia dubyi*. A, section of thallus edge × 325; B, section of mature thallus × 250; C–E, *Melobesia membranacea*. C, section of thallus edge × 600; D, section of older thallus × 600; E, section of older thallus × 730. From Dixon (1973) after Kylin and Rosenvinge by permission.

is not obviously of heterotrichous organization, with a prostrate system of filaments attached to the substrate by rhizoids and with upright filaments produced on the upper surface. Filaments equivalent to upright filaments are also produced on the lower surface. The thallus resembles an axis of an upright foliose type laid on its side.

According to Fritsch (1935), heterotrichy underlies the construction of many, if not all, of the more advanced red and brown seaweeds.

ADVANCED CONSTRUCTION

The formation of the large, structurally complex seaweeds takes place by the production of tissues made up of either pseudoparenchyma or parenchyma. Pseudoparenchymatous forms arise from the coalescence of filaments to produce a bulky thallus (Figure 22). In some forms the essentially filamentous nature of the pseudoparenchyma is easy to discern, but in advanced types the tissues superficially resemble

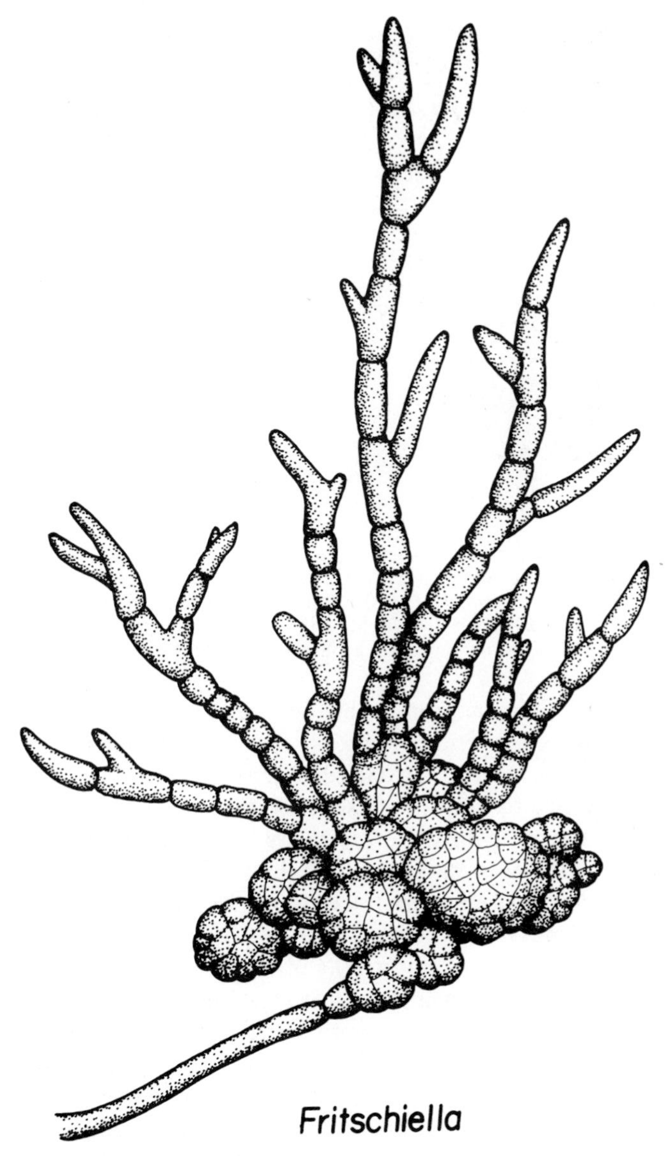

Figure 21. Stylized diagram of the heterotrichous green alga *Fritschiella* showing upright and prostrate portions of thallus. × 750

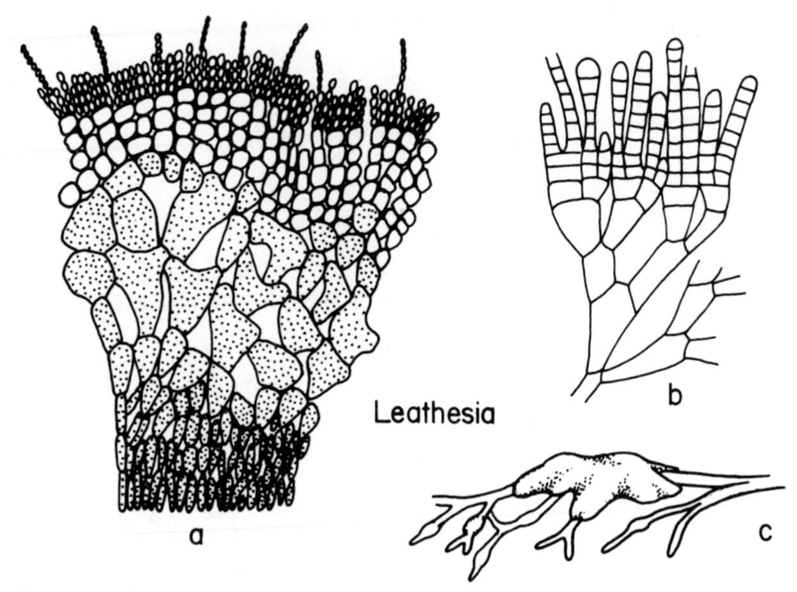

Figure 22. Multiaxial pseudoparenchyma in the brown alga *Leathesia*. a, cross section, × 135; b, stylized interpretation of cross section showing coalescence of branches; c, habit of plant growing on *Ascophyllum*, × 0.3. After Scagel et al. (1965), *An Evolutionary Survey of the Plant Kingdom*. Copyright © 1965 by Wadsworth Publishing Company, Inc., Belmont, Cal. By permission of the publisher.

parenchyma. There appear to be two general types of pseudoparenchymatous construction, uniaxial and multiaxial. In the former there is a single chain of axial cells around which lateral branches come to coalesce to form the false tissue (pseudoparenchyma) (Figure 23). In multiaxial thalli there are many axial cells series whose coalescence contributes to the bulk of the thallus (Figure 24).

The pseudoparenchymatous mode of construction is most elaborately expressed in the class Florideophyceae of the Rhodophyta. In these plants only the apical cells of the axial series divide at right angles to the axis of elongation. Each segment of the axial series gives rise by tangential divisions to pericentral cells, as Figure 25 shows. These pericentrals have three possible fates: a) to remain unchanged, b) to divide and form a cortex, and c) to become the apical cells of lateral branches of limited growth. The first fate is illustrated for a young plant of *Polysiphonia brodiae* in Figure 25A. The second fate, formation of a cortex, is illustrated in Figure 25B for *P. nigrescens*. Laterals of limited growth are shown in the simple construction of *Sirodotia suecica* in Figure 25C. In more complex forms there is considerable differentiation of cortex and medulla. In the leafy plants

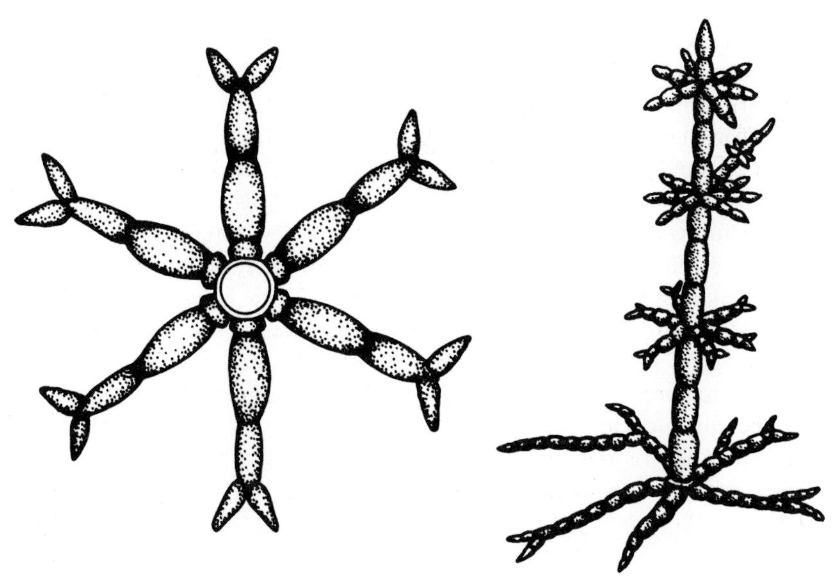

Figure 23. Stylized diagrams of uniaxial construction showing whorls of lateral branches. After Gayral (1975). By permission of Doin, Editeurs.

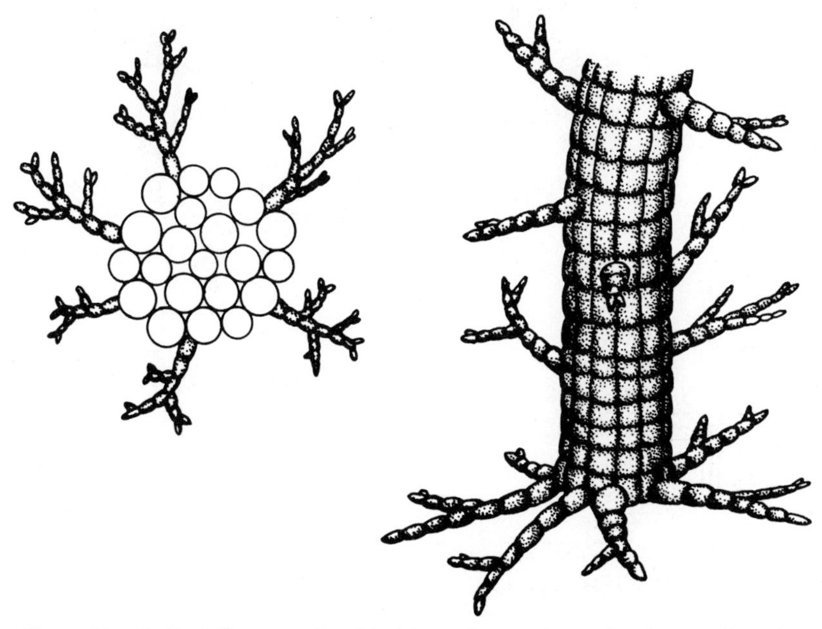

Figure 24. Stylized diagrams of multiaxial pseudoparenchyma showing a surface view (on right) and a cross section (on left). After Gayral (1975). By permission of Doin, Editeurs.

of the Delesseriaceae the broad flat thallus is produced by webbing of the laterals of limited growth. The best example of this type of construction is seen in *Hypoglossum* (Figure 26).

The largest seaweeds are of parenchymatous construction, in which a bulky thallus is produced by cell division in varying planes. This is shown very clearly in Figure 27 for *Sphacelaria*. The most elaborate parenchymatous forms are in the orders Fucales and Laminariales of the Phaeophyta. These plants show extensive differentiation of organs and tissues. In the kelps three tissue layers are formed: central medulla, inner cortex, and outer cortex (shown in part in Figure 28). The outermost layer consists of small actively dividing cells which are deeply pigmented, forming the assimilating layer. Cells of the inner cortex are larger and bear numerous cross connections and hyphae. In the medulla the elongate cells are spread out in a mucilage. These elongate cells seem to have a translocating role (see Chapter 4).

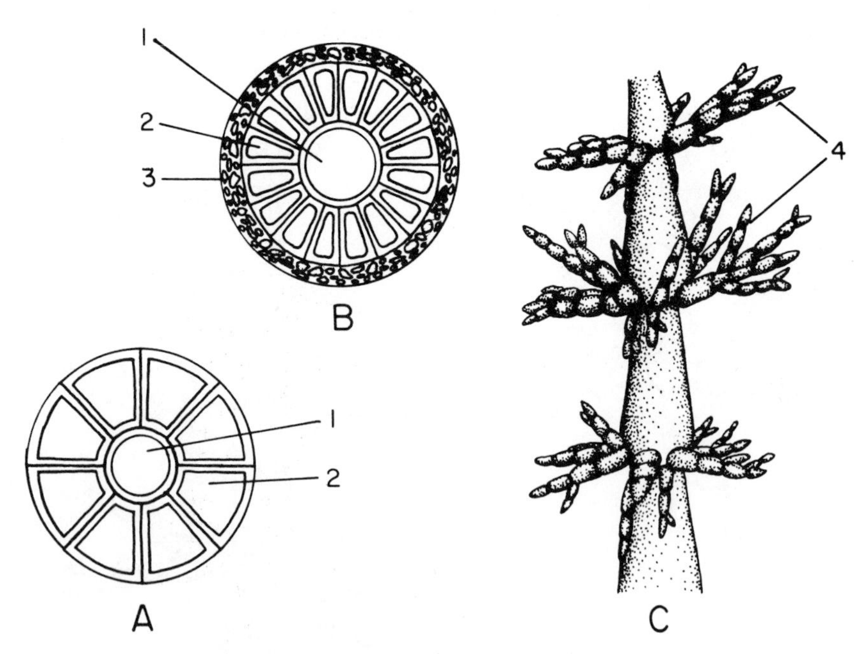

Figure 25. Stylized diagram showing the fates of pericentral initials. A, no further development; B, formation of a cortex; C, formation of laterals of limited growth. 1, axial cell; 2, pericentral cell; 3, cortex; 4, lateral of limited growth. A and B after Boney (1966), C after Dixon (1973).

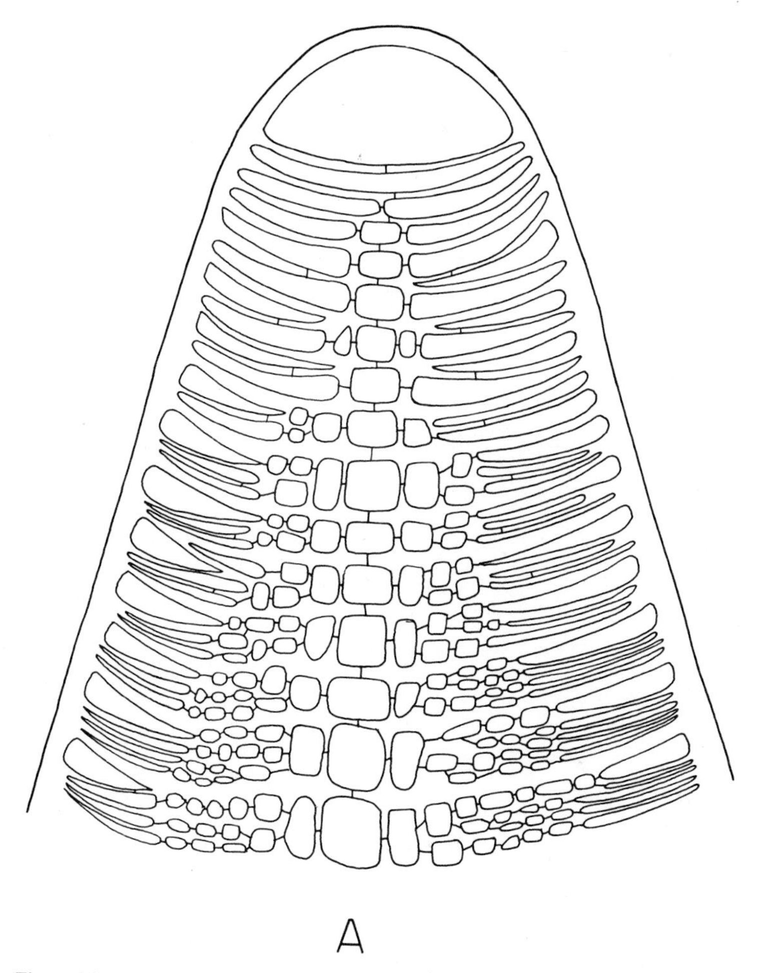

Figure 26. Apical organization in the red alga *Hypoglossum woodwardii*. × 1000.

The foliar differentiation of some of the Fucales is remarkable (Figure 29). In *Sargassum longifolium* the lateral branch systems have a distichous arrangement, and within each system there are two alternate "leaves," one with a prominent air bladder. Between the leaves the repeatedly branched reproductive shoots are arranged. In the Laminariales foliar organ differentiation is also well developed. This is shown especially in the genus *Egregia* (Figure 30), which has elaborate branching and the development of air bladders.

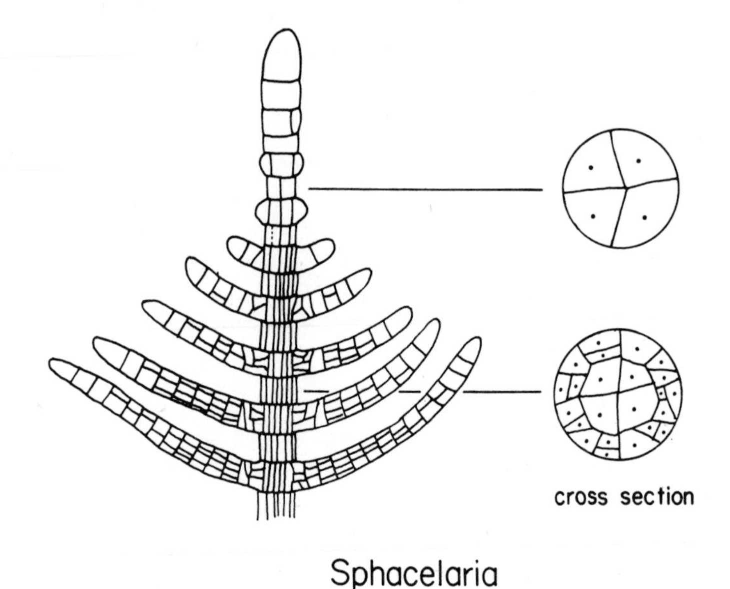

Figure 27. Stylized diagram to show the formation of true parenchyma in *Sphacelaria*. The apical cell acts as a meristem. After Scagel et al. (1965), *An Evolutionary Survey of the Plant Kingdom*. Copyright © by Wadsworth Publishing Company, Inc., Belmont, Cal. By permission of the publisher.

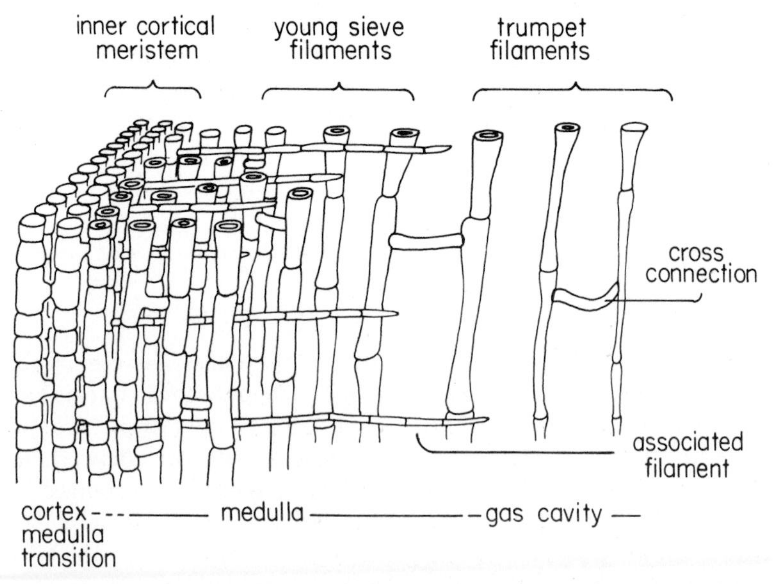

Figure 28. Stylized three-dimensional reconstruction of the medulla and part of the cortex in *Nereocystis*. After Nicholson (1976). *Botanica Marina XIX*: 28.

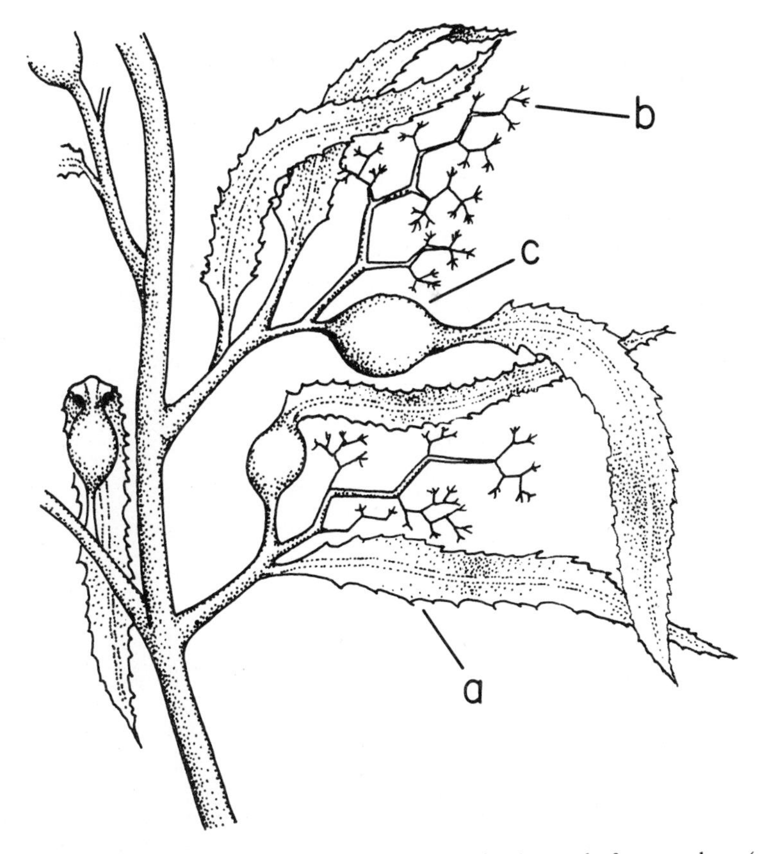

Figure 29. Foliar differentiation in *Sargassum* showing a leafy appendage (a), reproductive receptacles (b), and float (c).

From this account it should be clear that seaweeds are by no means simple forms of plants. They are, however, surpassed in complexity by vascular plants, which show a much greater differentiation of tissues and organs.

MERISTEMS

It is a general observation that as organisms increase in size and complexity there is a greater division of labor between the component parts. In the seaweeds one expression of this tendency is the development of meristems, localized regions of cell division and enlargement.

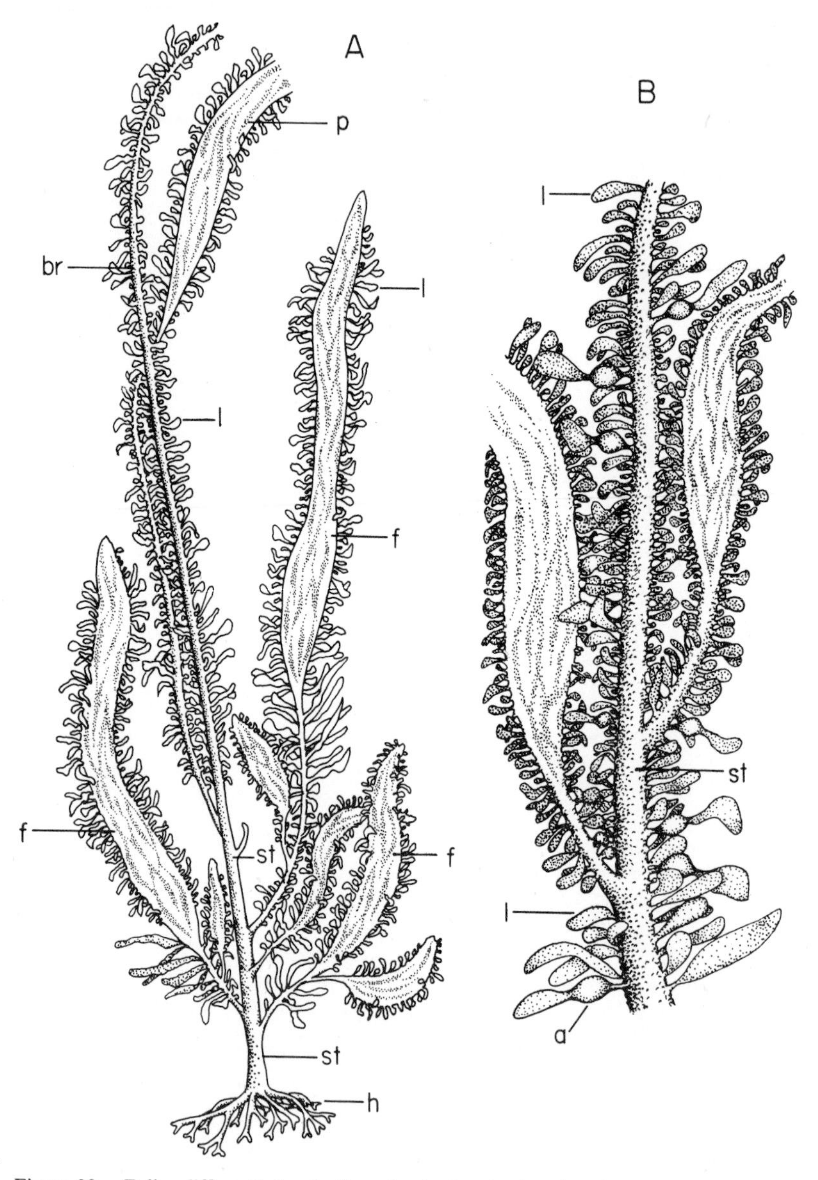

Figure 30. Foliar differentiation in *Egregia*. A, showing basal region; B, branch detail. a, air bladder; br, branch; f, flattened region of branch; h, holdfast; l, lateral; p, papillae; st, stipe. After Fritsch (1945) and Smith (1944). A × 0.1, B × 0.3.

The diversity of meristems is remarkable. In the simplest filaments any cell (except the basal, attaching cell) may undergo division. Growth is diffuse.

Meristems are of two types, intercalary and apical. Two types of intercalary meristems are found in the brown seaweeds, the trichothallic, and the much more massive form found in the Laminariales. A trichothallic meristem is situated at the base of a hair. Segments cut off contribute to elongation of the hair, but the majority are cut off below the hair (Figure 31). In the Laminariales the meristem is located in a transition zone between the foliar organs of blade and stipe.

Apical meristems consist of either a single cell or a group of apical initials. The former case is clearly shown in the genus

Figure 31. Trichothallic meristem (bracketed) of a 47-day-old sporophyte of *Desmarestia aculeata*. × 160.

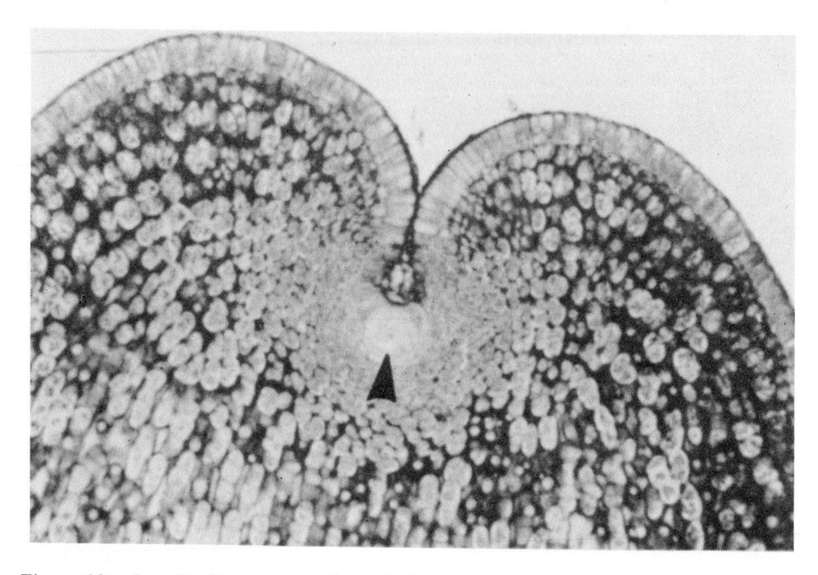

Figure 32. Longitudinal section through the apex of *Ascophyllum* showing the large apical cell (arrowed) and surrounding meristematic cells. From Moss (1974). By permission of Blackwell Scientific Publications Ltd.

Sphacelaria, depicted in Figure 27. In this case all of the cells of the thallus originate from the apical initial. Until recently it was supposed that in the Fucaceae the cells of the thallus were derived from a single initial. In 1967, Moss demonstrated that the apical cell of this genus divides only rarely and that the cells around it display meristematic activity (Figure 32).

Morphogenetic aspects of meristems are considered in the next chapter.

Chapter 4
Functioning of Whole Plants

Much is known of the structure of seaweeds. Less is known of their physiological functions. This chapter is concerned with the vegetative functioning of whole plants as opposed to their constituent cells. Reproductive functions are considered in the next chapter.

The first three sections of this chapter deal with the responses of plants to their physicochemical environment. Essentially this is physiological ecology and contrasts with the material in the last two sections, which is concerned with the intrinsic functions of translocation, storage, and development.

WATER MOVEMENT AND GROWTH FORM

The previous chapter related all of the diverse structures of seaweeds through their development patterns. Thus heterotrichy seems to be a modification of the filamentous type of construction. Pseudoparenchyma is derived from a coalescence of filaments. Parenchyma arises by cell division in many planes. The end result of these developmental patterns is the growth form of the adult plant.

According to Neushul (1972), seaweeds may occupy one or more regions of water motion. Three of these water motion regions are shown in Figure 33. In the current zone of a kelp forest the water moves in one direction at an approximate velocity of 1 $m^{-1}sec^{-1}$. Closer to the bottom the plants are seen to move backward and forward in the surge zone. Here there is an oscillating water motion produced by waves, as Figure 34 shows. In the lowermost region of water motion, the boundary layer (about 2 cm thick), water movement is much slower. If the water is oscillating at about 1 $m^{-1}sec^{-1}$ above the bottom, then friction at the boundary layer will reduce the velocity to one-tenth or less of this. Thus reduced flow of water severely limits nutrient availability at this level.

Neushul has shown that the growth forms occupying the boundary layer can be divided into three main categories: a) encrusting, b) filamentous, or c) thin bladed. The encrusting or pad-like forms may represent a means of competing for space. On the other hand, filamentous construction provides for better nutrient absorption since the thallus projects up into the faster moving water.

Charters, Neushul, and Barilotti (1969) have studied the movement of the kelp *Eisenia arborea* in the surge zone. Figure 35 shows the morphology of this plant. As water flow velocity increases over an *Eisenia* plant the blades and stipe bend more in the flow direction. This bending of the stipe brings the crown of blades down nearer to the substate where baffling by other organisms and rocks causes decreased flow velocity. By bending with the flow, the blades reduce their projected area exposed at high angles to the flow. Thus form drag is decreased by a reduction in both velocity and projected area.

Plants that extend up into the current zone are equipped with floats which hold up photosynthetic blades near the surface. This is

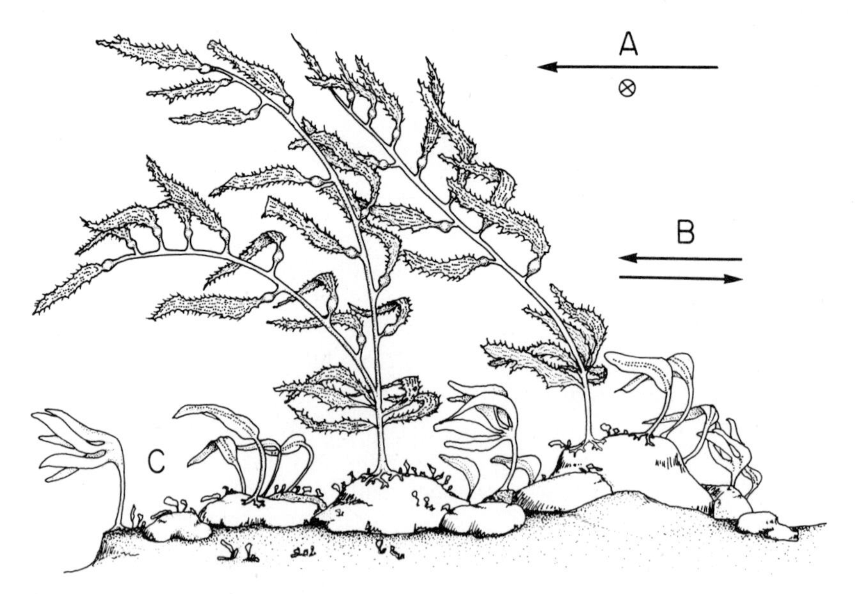

Figure 33. Water motion regions in a California kelp forest. A, current zone; B, surge zone; C, boundary layer. After Neushul (1972).

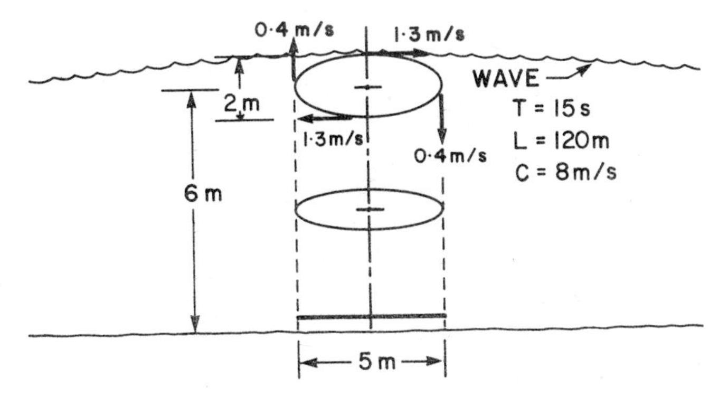

Figure 34. The movement of water in a surge zone when a wave motion sweeps the bottom. A 2-m wave moved past at 8 m/sec^{-1} (C) with a wavelength (L) of 120 m and a wave period (T) of 15 sec to produce orbital velocities (arrows) shown. These, near the bottom, flattened to produce the back and forth water motion of the surge zone. From Neushul (1972).

shown for *Macrocystis* in Figure 33. The long stipes are very flexible and commonly coiled. They serve to dampen wave forces transmitted to the holdfast. The force transmitted down a single stipe bundle may be more than 8 kg. The blades of the plants in the current zone are themselves subject to the formation of a boundary layer. Thus we often see complicated surface corrugations which enhance surface turbulence and nutrient uptake. There are similar surface configurations in some kelps in the lower surge zone. *Agarum cribosum* has blades that are perforated with numerous holes, each of which has a raised lip that serves to increase surface turbulence.

RESPONSES OF PLANTS TO LIGHT

Plants respond to light through their photosynthetic activity, which clearly influences growth rates. There are other, nonphotosynthetic responses, which are conveniently termed photomorphogenic (Dring, 1971).

Seaweeds differ in their pigmentation and are thus known as red, green, or brown algae. The light made available to them may also vary in its color quality. As light penetrates coastal waters, there is a logarithmic reduction in intensity and a selective absorption of various spectral components (Figure 36). The photic zone is generally that region of the water column in which irradiance is $\geq$ 1% of surface value (Dring, 1971). The depth of this zone obviously depends on the

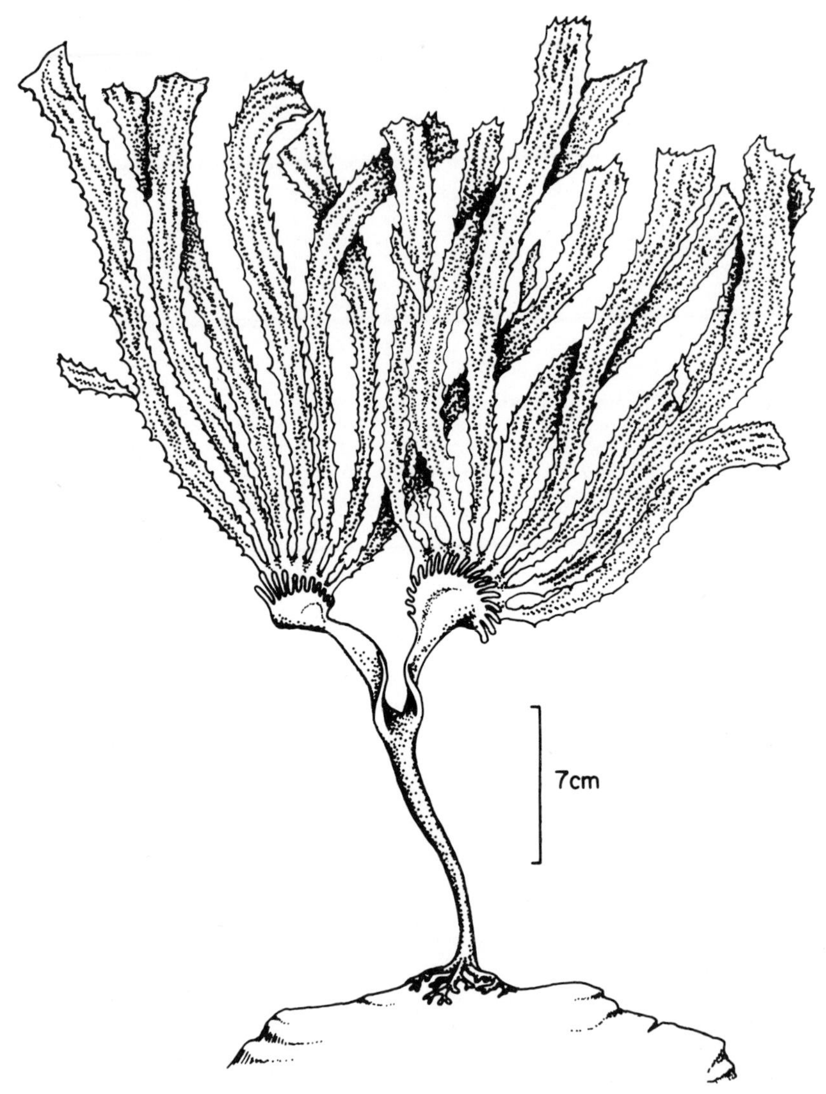

Figure 35. Morphology of the kelp *Eisenia arborea* showing the crown of sporophylls. After Scagel (1967).

light attenuation properties of the water. In the lower reaches of the photic zone in coastal waters most of the light is in the blue-green region of the visible wavelengths. Algae of different colors have different abilities in absorbing light of various wavelengths, as shown in Figure 37. They also differ in their utilization of absorbed light, and this is expressed in the action spectrum, which relates relative photosynthetic rates to light color. Generally, green algae utilize red and blue light most and green light least. Brown algae also use red and blue, but have a wider extension into the blue-green region. Red algae make maximum use of green, less of red, and least of blue.

If irradiance is I_x and photosynthetic rate P at wavelength λ, then the total value of photosynthesis Q at a given depth (x) can be calculated planimetrically:

$$Q_x = \int P(\lambda) \cdot I_x(\lambda) \cdot d\lambda$$

This theoretically means that as depth increases the photosynthetic activity of green algae is inhibited more quickly than that of red or brown algae. Levring (1947) has shown this to be so in experiments he carried out in the sea off the coast of Sweden. This work is put forward as supporting evidence for Engelmann's (1884) theory of complementary chromatic adaption, which states that the color of algae is complementary to the prevailing light by which they are illuminated. This theory and the experimental evidence help explain

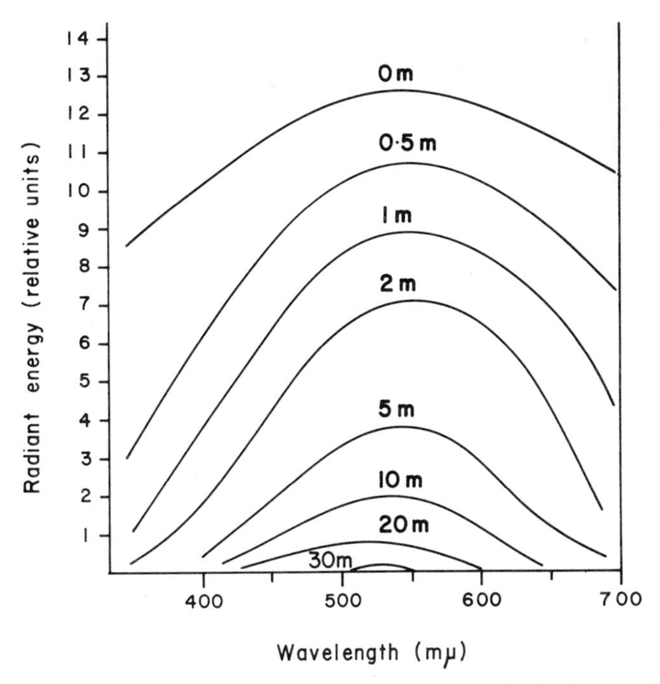

Figure 36. Spectral distribution of light energy at different depths in normal coastal water. From Levring (1947) by permission.

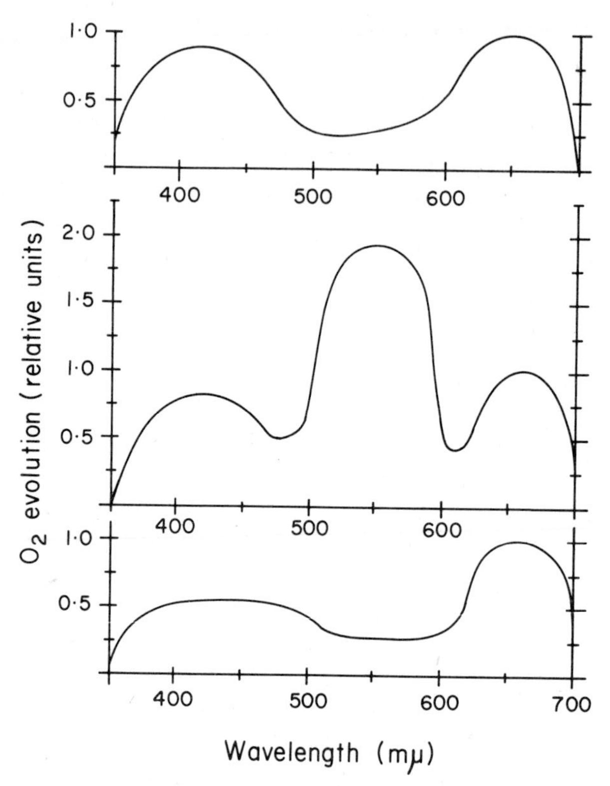

Figure 37. Photosynthetic rates of green, red, and brown seaweeds under various light wavelengths. The upper curve is for *Ulva lactuca* (Chlorophyta), the middle curve for *Delesseria sanguinea* (Rholdophyta), and the lower curve for *Fucus vesiculosus* (Phaeophyta). From Levring (1947) by permission.

why green algae tend to occur mainly between the tide marks, brown algae in the midshore and sublittoral, and most red algae in deeper water (Table 5).

Table 5 clearly shows that some green algae are able to grow in the sublittoral. In fact, off the coast of South Africa, species of the genus *Codium* occur in waters of considerable depth. Ramus, Beale, and Mauzerall (1976a) and Ramus et al. (1976b) have thus reexamined the whole question of chromatic adaptation. They found that membranous forms such as *Porphyra* (red alga) and *Ulva* (green alga) did indeed show a differential photosynthetic response to light quality at different depths. This contrasts markedly with the fleshy thalloid forms *Chondrus* (red alga) and *Codium* (green alga). *Codium* can perform at least as well as *Chondrus* in deeper waters. The reason for the different responses in membranous and thalloid forms is related to their differing abilities to absorb light. In *Ulva* and *Porphyra* most of the incident light is not absorbed. However *Codium* and *Chondrus* absorb most of the incident light and are thus optically black even though they may appear green and red on the surface. This is confirmed subjectively by any diver in deep water, who will notice that most of the seaweeds appear black. If all of the incident light is absorbed, differential abilities in absorbing light of differing wavelengths becomes irrelevant.

This discussion takes no account of the fact that individual plants have the ability to change: a) their total pigment concentration and b) the proportions of their various pigments. Ramus et al. (1976a) found that intertidal plants occur in sun and shade forms that differ in their total pigment concentration but not their pigment proportions. On the other hand, subtidal forms change their pigment proportions within 7 days when moved to a depth of differing spectral light quality. Pigment concentrations also change in the subtidal plants.

The responses of plants to light discussed previously are all related to their photosynthetic pigments. There are other nonphotosynthetic pigments which differ from the photosynthetic pigments in their absorption spectra and energy requirements. Dring (1971) groups the nonphotosynthetic responses of seaweeds to light as: a) phytochrome-controlled responses, b) photomorphogenic responses of *Acetabularia*, and c) photomorphogenic responses to blue light.

a. In seaweeds the phytochrome-controlled response is manifested in the phenomenon of photoperiodism. Photoperiodism is characterized by the induction or continuation of a response following transference to noninducing conditions after a set time and sensitivity to short breaks in either light or dark periods. A clear example of photoperiodism is seen in the work of Powell (1964;

Table 5. Vertical distribution of Chlorophyta, Phaeophyta, and Rhodophyta species on the island of Colonsay

	Percent of species		
Location	Chlorophyta	Phaeophyta	Rhodophyta
Littoral only	57	30	15.2
Littoral and sublittoral	28.5	30	18.5
Sublittoral only	14.5	40	66.3

From data in Norton, McAllister, and Conway (1969).

cited in Dixon, 1973) on *Constantinea subulifera*. In this red alga initiation of the blade is induced by a critical day length of 14 hr. In longer day lengths development is inhibited. A 15-min light period in short day conditions also inhibits development. In the red algae *Porphyra* and *Bangia* regulation of spore formation and development is under photoperiodic control (Dring, 1967; Richardson and Dixon, 1968). Photoperiodic responses of this type are usually associated with a pigment known as phytochrome which, in higher plants, acts as follows. Phytochrome exists in two interconvertible forms, R-phytochrome and F-phytochrome. The interconversion depends on light quality:

$$\underset{P_{660}}{\text{R-phytochrome}} \underset{\substack{\text{far red} \\ \text{700–760 nm}}}{\overset{\substack{\text{orange-red} \\ \text{600–680 nm}}}{\rightleftarrows}} \underset{P_{730}}{\text{F-phytochrome}}$$

metabolically in darkness

When the pigment is in the R-phytochrome form the plant's biochemistry is adjusted so that it is typical of a plant in the dark. F-phytochrome is converted metabolically to R-phytochrome. In this way plants are able to determine with great precision the duration of dark periods. Photoperiodism mediated by blue light is discussed in (c).

b. Photomorphogenic responses of *Acetabularia*. *A. crenulata* and *A. mediterranea* plants develop caps in blue or white light but not in red light (Terborgh, 1965). There is no action spectrum available for these responses, so the nature of the photoreceptor is unknown.

c. Photomorphogenic responses to blue light. The responses of seaweeds to blue light are varied, and it is not known if there is a single underlying pigment photoreceptor equivalent to phytochrome. Lüning and Dring (1975) have investigated the effects of blue light on gametogenesis in the female gametophytes of *Laminaria*. Figure 38 shows the action spectrum of the response. It differs markedly from the photosynthetic action spectrum, which peaks at 670–680 nm. Blue light also initiates two-dimensional growth and hair formation in the brown alga *Scytosiphon lomentaria* (Dring and Lüning, 1975). They also showed that the crustose phase of *Scytosiphon lomentaria* exhibits a blue light-mediated photoperiodic response. In 16 hr of white light per day

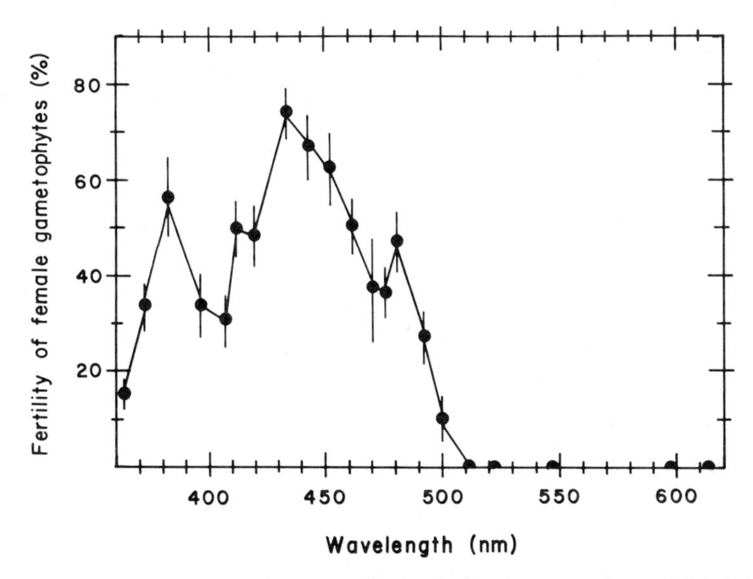

Figure 38. Action spectrum for egg production in female gametophytes of *Laminaria saccharina*. Values shown are means ± 95% confidence limits. From Lüning and Dring (1975).

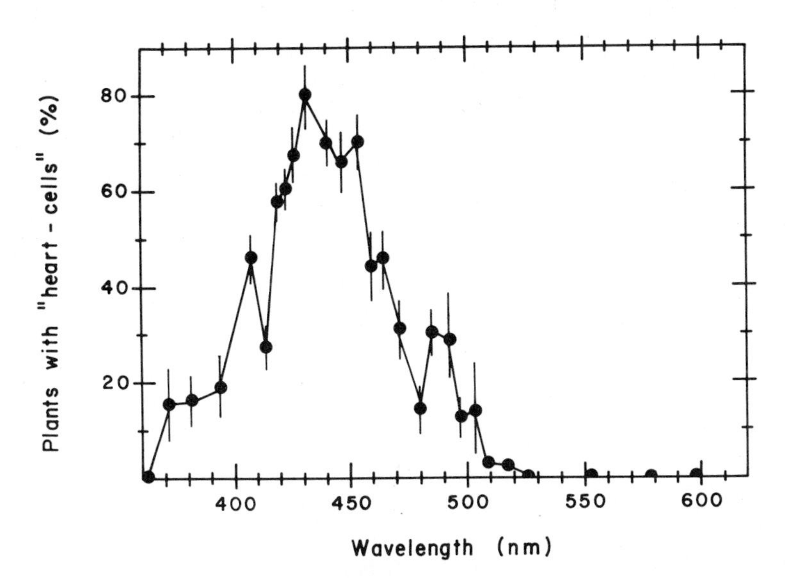

Figure 39. Action spectrum for the induction of two-dimensional growth in *Scytosiphon lomentaria*. Two-dimensional growth is initiated by the development of "heart-shaped" apical cells. Counts were made of these and are expressed as means ± 95% confidence limits. From Dring and Lüning (1975).

the crustose phase persisted indefinitely. Upon transfer to day lengths of less than 12 hr, erect cylindrical thalli were produced. A 1-min light break of low blue irradiance in the middle of a 16-hr night inhibited the short day response. Figure 39 shows the action spectrum for the response (Dring and Lüning, 1975). Red light is without effect in this response. In this respect *Scytosiphon* is unlike all other short day plants investigated.

In *Nitophyllum punctatum* tetraspore discharge occurs rhythmically under daily light-dark periodicity. Blue light enhances tetraspore formation (Sagromsky, 1961).

The phenomenon of phototropism is known among seaweeds, and again response is greatest in blue light and least in red. As early as 1882 Berthold reported that in *Bryopsis* tips of branches grow toward light and rhizoids away. It seems likely that in all of these cases carotenoids or flavins are the photoreceptors.

RESPONSES OF PLANTS TO SALINITY

The salt content of seawater is reduced by influx of fresh water and raised by evaporation. Tolerance of these variations is an important functional response of marine plants. It is well known that species from intertidal environments have much greater tolerance to osmotic change than those from the sublittoral (Table 6). Algae that are resistant to osmotic shock are termed *euryhaline*. Those that are not resistant are termed *stenohaline*.

Changes in salinity affect the osmotic relationship of algal cells and the medium in which they are bathed. The process by which an

Table 6. Range of salinities tolerated by seaweeds from intertidal and subtidal locations

Species	Location	Range of salinities tolerated (1 = full seawater)
Polysiphonia urceolata	Intertidal	0.3–2.0
Membranoptera alata	Intertidal	0.4–1.9
Ptilota plumosa	Intertidal	0.4–2.2
Ceramium ciliatum	Intertidal	0.4–2.2
Heterosiphonia plumosa	Subtidal	0.6–1.3
Cryptopleura ramosa	Subtidal	0.5–1.4
Brongniartella byssoides	Subtidal	0.4–1.4
Phycodrys rubens	Subtidal	0.6–2.0

From Gessner and Schramm (1971).

alga overcomes osmotic shock is termed *osmoregulation*. This topic has been reviewed recently by Hellebust (1976). Cells of euryhaline algae respond to external osmotic changes by altering the internal concentrations of organic solutes or inorganic ions. Kauss (1968) has demonstrated that the synthesis of floridoside and isofloridoside (α-galactosyl glycerols) in *Porphyra perforata* is stimulated in a linear fashion by increasing the salt content of the surrounding medium. Presumably this action increases the internal osmotic potential and prevents cellular plasmolysis. Eppley and Cyrus (1960) showed that in *P. perforata* intracellular K^+ increases with increased salinity and this also serves in osmoregulation. The K^+ ion pump osmoregulator has been extensively studied in the coenocytic genus *Valonia*. If the internal turgor pressure of cells is decreased there is a marked influx of K^+ ions. *Valonia* therefore, appears to have a feedback mechanism for controlling turgor pressure. Hellebust (1976) visualizes a pressure-sensitive transducer located in the plasmalemma or plasmalemma wall complex.

INTERNAL TRANSPORT

Subtidal seaweeds are continuously bathed in water. Therefore they have no requirement for an internal water transport mechanism equivalent to the xylem of vascular plants. Intertidal seaweeds are exposed to air periodically, but there is no evidence of a water transport system in these plants either. Like vascular plants, many of the structurally complicated seaweeds have source-sink relationships in the production of photosynthetic carbohydrates. For instance, in the kelp *Laminaria* the blades may form a canopy that shades out the lowermost portions of the plant. The blade with a large surface area traps most of the light and carries out most of the photosynthetic activity of the plant. The blade is a photosynthetic source, and the lowermost portions form a sink. The meristem at the junction of the blade and stipe requires large quantities of carbohydrate for the formation of new tissues. These observations lead to the hypothesis that there may be translocation of the products of photosynthesis. This has been shown to be true in a wide variety of kelp species surveyed by Schmitz and Lobban (1976). The technique used for tracing movement of translocate was straightforward. A Plexiglas feeding chamber was glued to the blade surface (cyanoacrylate adhesive). ^{14}C-Labeled bicarbonate dissolved in seawater was pumped through the chamber. The labeled bicarbonate ions were absorbed by the photosynthesizing tissues and converted into organic products. Subsequently, the movement of the radioactively labeled products was followed by the use of autoradiography. In this technique sensitive film is applied to the plant and radioactive emissions cause the formation of an image (Figure 40). The translocation rates in kelps are in the range 50–780 cm/hr, which is very similar to vascular plants.

There are two important questions concerning translocation in seaweeds: In which tissues and cells does movement of assimilate occur? What is the translocate? The question of which tissues translocate has been examined by Steinbiss and Schmitz (1973) and Schmitz and Srivastava (1975). As pointed out in the last chapter the medulla of laminarians contains elongate elements known as sieve tubes or sieve elements (Figure 41). The cross walls between joined sieve tubes are perforated by numerous sieve pores through which the protoplasts are interconnected. Microautoradiography shows that the sieve tubes are conducting elements in *Laminaria hyperborea* (Steinbiss and Schmitz, 1973). This is almost certainly the route in other brown algae.

Schmitz, Lüning, and Willenbrink (1972) and Schmitz and Srivastava (1975) have shown that the translocate in kelps consists of the polyol mannitol and a variety of amino acids. According to Schmitz and Srivastava (1975), *Alaria marginata* may have little selectivity in translocation. Whatever substances are produced photosynthetically are translocated.

There is some controversy at present over whether or not transport of phosphorus (as ^{32}P) occurs in kelps. Floc'h and Penot (1972) report that when various laminarian species are fed $H_3\,^{32}PO_4$ there is a transport back to the meristem and the holdfast. However

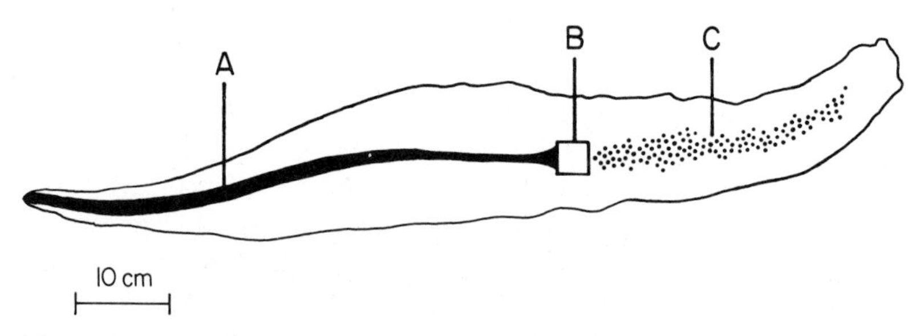

Figure 40. Translocation in a *Laminaria saccharina* blade. The sketch shows an autoradiograph after application of $H^{14}CO_3^-$ to the tissue marked B. Trace A shows the path of radioactive assimilate back to the meristem. C shows position of sorus.

Figure 41. Radial longitudinal section through the midrib at the intercalary growing region of *Alaria marginata*. The inner cortex cells (C) are arranged in vertical files which separate from each other along their longitudinal walls. In the medulla the sieve tubes (S) and hyphal cells (H) seem embedded in a matrix of alginic acid. From Schmitz and Srivastava (1975). Reproduced by permission of the National Research Council of Canada from the *Canadian Journal of Botany 53*: 861–876.

Schmitz and Srivastava (1975) were unable to demonstrate any such transport in *Alaria marginata*. They also argue that since algae are able to absorb inorganic ions over their entire surface, it would be difficult to assign a functional role to long distance transport. However, it should be pointed out that, in the sea, inorganic ions such as PO_4^- and NO_3^- are present in extremely low concentrations. Although all cells may receive ions by diffusion from the thallus surface, this may be insufficient for rapid growth at the meristem. It may be that the meristem receives a supplement to the diffusion supply by long distance transport. The work of Chapman and Craigie (1977) is relevant here. They showed that *Laminaria longicruris* accumulates NO_3^- ions during the winter to concentrations that are 28,000 times higher than the surrounding water. These internal reserves are present in the tissues for 6–8 weeks after the depletion of dissolved NO_3^- in the water column in the spring. During this 6–8 weeks growth is maintained at a high rate. It seems possible that the nitrate reserves are supplied to the meristem. This needs testing experimentally. It may be that the nitrogen is transported as amino acids, which are already known to be components of the translocate (Schmitz and Srivastava, 1975).

GROWTH AND DIFFERENTIATION OF WHOLE PLANTS

In the Florideophyceae many species are built up on a distinct hierarchical system, with the growth of the primary axis greater than that of the secondary, growth of the secondary axis greater than that of the tertiary, and so on. Dixon (1973) assumes that there is hormonal control over this hierarchical organization, but little information is available in this regard. In forms that are differentiated into laterals of limited and unlimited growth (see previous section, "Advanced Construction"), the process of limitation can be disrupted if the apex of a filament of unlimited growth is surgically removed. This suggests that a morphogenetic substance travels from the apex of a filament of unlimited growth to the more basal regions.

Observations on polarity in algae also point to the occurrence of morphogenetic substances or hormones. Most of the work in this field has been done on *Fucus* zygotes. A fertilized egg begins its development by the formation of an attaching rhizoid on one side. The first division of the egg is transverse to the rhizoid and so determines the polarity of cell division. The morphological polarity is probably a manifestation of a polarity gradient of chemical substances. The experimental evidence certainly supports this view. Thus the polarity of *Fucus* eggs is affected by the direction of illumination, electric current, pH gradient, presence of other eggs, and centrifugation. These effects may be related to the distribution of cytoplasmic RNA in the egg (Nakazawa, 1975). Before the rhizoid is formed RNA is uniformly distributed. After swelling of the rhizoid the RNA is transported to the tip and takes part in rhizoid elongation. If RNA synthesis is inhibited by actinomycin D, rhizoidal protuberances do not form. RNA molecules are negatively charged, and therefore response to an electric field is expected. Similarly, eggs placed in a potassium ion gradient develop rhizoids on the higher side of the gradient. This might be accounted for by the migration of negatively charged RNA toward the pole of higher concentration of positive ions. The evidence available points clearly to the occurrence of gradients of morphogenetic substances or hormones in seaweeds. However there is no consensus as to the chemical nature of these substances. According to Buggeln (1976), Buggeln and Craigie (1971), and Provasoli and

Carlucci (1974), there is no unequivocal evidence that identifies known land plant hormones as growth substances in algae. According to Provasoli and Carlucci (1974), two procedures have been used in attempting to determine the presence of hormones: 1) extraction, identification by chromatography, and plant bioassay (coleoptile test); and 2) application of growth substances experimentally and observations on the physiological responses.

Very careful thin layer chromatography has failed to identify endogenous bound IAA (Buggeln and Craigie, 1971) in a range of seaweed species. Similarly Dawes (1971) was unable to detect IAA in *Caulerpa prolifera* but was able to obtain growth modification by exogenous application of the hormone. This brings into question experiments that attempt to identify the presence of auxin by exogenous application and subsequent growth modification. Growth modification can be achieved by the application of organic compounds that do not occur naturally in plants. These compounds are often structurally related to naturally occurring higher plant auxins. Buggeln (1976) has carried out critical experiments to test whether or not the kelp *Alaria esculenta* can differentiate (through its growth response) between an auxin and its related nonauxin (eg., (2,4,-

Figure 42. Development of organized primordia (arrow) in shaker culture of juvenile phase of *Codium*. From Ramus (1972), *American Journal of Botany 59*: 481.

dichlorophenoxy- and 3,5-dichlorophenoxyacetic acids). There was no distinction between the effects of five pair of auxin and nonauxin compounds.

Evidence for the occurrence of cytokinins and gibberellins is similarly suspect. This means that although we can be sure that hormones occur in algae, their identity is unknown. Furthermore, work by Provasoli and Pintner (1964) indicates that many seaweeds require exogenous growth substances for normal development. For example, *Monostroma oxyspermun*, a green alga of parenchymatous construction, loses its normal morphology in bacteria-free culture. Only long rhizoidal cells and aggregates of one to three rounded cells develop. Normal morphology was not restored by the application of known plant hormones or nutrients. However, the filter-sterilized supernatants of two marine bacterial isolates and several bacteria-free red algae and the brown substances released from *Sphacelaria* or *Fucus* induced normal development.

Besides requiring growth substances for normal development, some algae also need appropriate physical conditions. This is shown clearly in the pseudoparenchymatous siphonous genus *Codium*, which breaks down into an undifferentiated mass of filaments in still culture. Ramus (1972) demonstrated that by applying a shear force on a shaker, normal filament aggregation was induced (Figure 42).

It is surprising that so little is known of the intrinsic aspects of morphogenesis in seaweeds in comparison with vascular plants. Embryogenesis is much simpler in seaweeds and the meristems of many forms are much larger and easier to work with than those of higher plants.

Chapter 5
Reproduction

Reproduction and life histories in seaweed are fascinating, if poorly understood, topics. This chapter examines vegetative propagation, asexual reproduction, sexual reproduction, life histories, and physiological processes. Under physiological processes environmental effects, reproductive hormones, and endogenous rhythms are considered.

VEGETATIVE AND ASEXUAL REPRODUCTION

Vegetative reproduction involves the separation from a parent plant of a vegetative component that later develops into a new individual. At least some of the cell walls of the original parent are incorporated into the new plant. The way in which separation occurs may be by simple fragmentation, as in *Ectocarpus* (Russell, 1967). In another brown alga, *Sphacelaria*, distinctive organs of vegetative propagation are formed. These are the propagules (Figure 43). Propagules are easily dislodged by vigorous water movement. After settlement they soon develop attaching rhizoids from the arms touching the substrate.

Asexual reproduction in seaweeds differs from vegetative propagation in that it involves the formation of special reproductive spores. These spores are either without cell walls or they may form their own walls. In either case the cell walls of the parent plant are not incorporated in the cells of its offspring. In almost all green and brown seaweeds asexual spores are zoospores. These are unicellular flagellated bodies (Figure 44) formed within sporangia. In the simplest filamentous green algae any cell of the thallus, except the holdfast, may produce zoospores. Sporangia are not morphologically differentiated from ordinary vegetative cells. In the brown algae many forms have well-differentiated sporangia, such as the pluriocular structures in *Ectocarpus* (Figure 45).

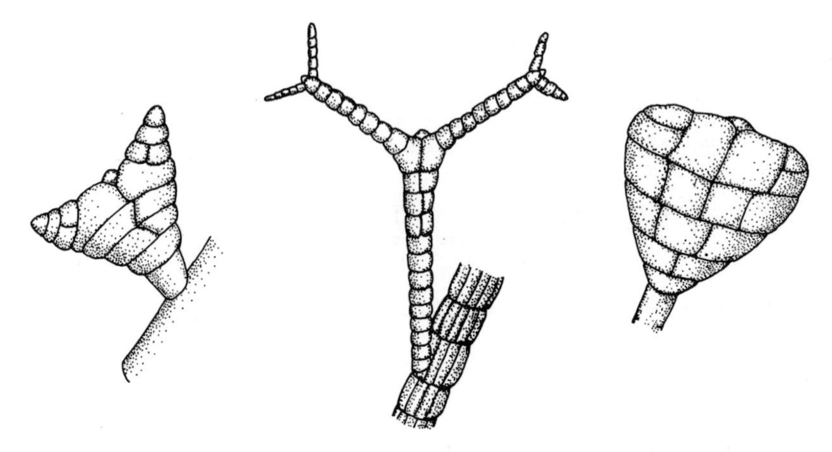

Figure 43. Various types of vegetative propagules found in *Sphacelaria*. After Dawson (1966).

Sporangia of the advanced brown algae are sometimes aggregated in macroscopic fruiting structures known as sori. Sori are easily visible as darkened raised areas on the blades of laminarians. In some of families of the Laminariales (eg., Alariaceae) the sori are born on special reproductive blades known as sporophylls (Figure 46).

In the red algae all of the reproductive bodies, including asexual spores, are nonmotile. There are many structurally different sporangia and means of spore production in red algae (see Dixon, 1973), but they all play a role in asexual reproduction. In none of the seaweeds do the spores act as resting stages. In this respect they differ markedly from many freshwater algae in which the reproductive bodies form thick-walled cysts or resting stages for the survival of unfavorable environmental conditions. The more stable conditions in the sea may explain the difference.

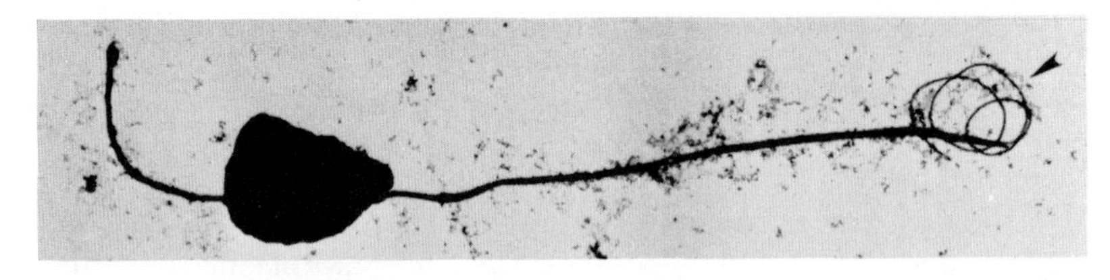

Figure 44. Whole mount of a zoospore of *Chorda* showing a long anterior flagellum with a tightly coiled distal portion (arrow). × 3250. From Toth (1976). Courtesy of R. Toth and the *Journal of Phycology*.

Figure 45. Many-chambered plurilocular sporangium in *Ectocarpus*.

SEXUAL REPRODUCTION

Sexual reproduction involves the process of gamete fusion or syngamy. Karyogamy or nuclear fusion leads to a doubling of the chromosome complement. Therefore, the process of meiosis is required to reestablish the haploid condition. Gamete fusion in some green seaweeds (eg., *Cladophora*) is between bodies of similar morphology. These are known as isogametes, and the sexual process as isogamy. Isogametes are motile, flagellated structures and fertilization is effected in the planktonic state. Anisogametes are also motile, but one is larger and less motile than the other. All of the advanced seaweeds have oogamy, in which the female gamete (egg) is nonmotile and considerably larger than the sperm. The egg may be shed before fertilization or retained on the parent plant.

The reproductive structures that develop gametes are termed gametangia. The degree of gametangial differentiation varies among seaweeds in the same way that sporangial differentiation varies. Only in some of the brown algae are multicellular gametangia formed. In the Florideophyceae (red algae) the female gametangium is called a carpogonium. Its distinctive shape is shown in Figure 47. The elongate neck is termed a trichogyne and it functions as a receptor for the male

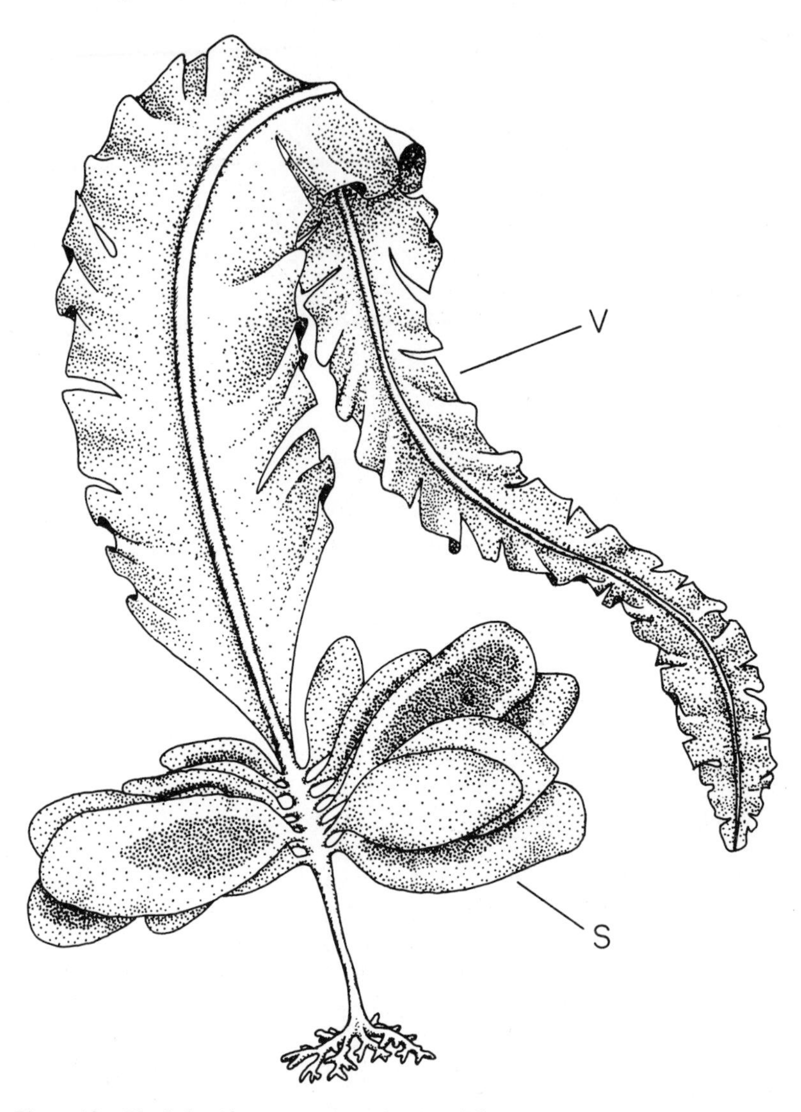

Figure 46. The kelp *Alaria marginata* showing differentiation into vegetative (V) and reproductive blades (S). × 0.3. After Smith (1944).

gametes, which are known as spermatia. The bulbous lower part of the carpogonium contains the egg nucleus.

The process of sexual mating has been studied most extensively in algae with motile male and female gametes. In most species studied the process proceeds through the following stages: a) gamete formation and release, b) clumping of gametes, c) pairing, and d) cell fusion. When released gametes of appropriate mating strains are mixed, an agglutination or clumping occurs. The site of agglutination is the flagellum tip in either one sex (brown algae) or both sexes (green algae). After clumping, pairs of attached gametes break away from the clumps. After pairing, cell fusion occurs.

When male and female gametes are born on the same plant there may be self-compatability or self-sterility. Apart from self-sterility, the occurrence of distinct mating types is common in seaweeds. These are male (+) or female (−). This type of mating system is termed *bipolar*. Within a single species there may be populations that are reproductively isolated. Each of the sexually isolated groups contains one pair of mating types. This has been demonstrated most clearly by Müller (1976b) for the brown alga *Ectocarpus siliculosus*. Müller (1968) also showed that sex determination in *Ectocarpus* is genotypic, with segregation taking place during meiosis. In the Laminariales, sex determination is related to the occurrence of sex chromosomes (Evans, 1965).

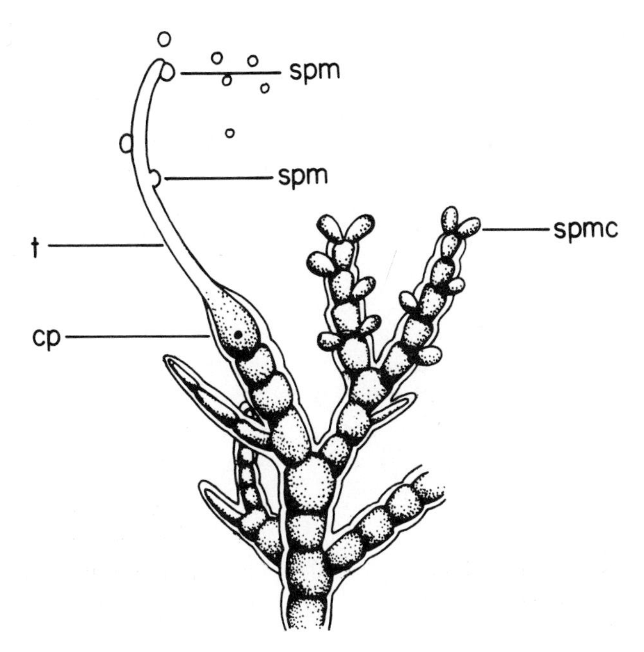

Figure 47. Gametangia in the red alga *Nemalion*. The female gametangium (cp, carpogonium) has an extended neck (t, trichogyne) which acts as a receptor for the male gametes (spm, spermatium) which are produced in a spermatangium (spmc). After Gayral (1975). By permission of Doin, Editeurs.

The existence of relative sexuality as described by Hartmann (1925) is now entirely discounted (Müller, 1976a). Reference to this phenomenon in earlier phycological works should be ignored.

PHYSIOLOGY OF REPRODUCTION

Hormones

The hormones involved in the mating reaction of microscopic algae have been studied more extensively than those of seaweeds. Mating hormones have been identified in the following processes in seaweeds: release of gametes, chemotaxis, and clumping. Lüning (personal communication) has recently demonstrated that cell-free culture medium supernatant of *Laminaria* female gametophytes triggers the release of sperms from antheridia of male gametophytes. The gamone (reproductive hormone) is not species specific.

There is considerable evidence that female (−ve) gametes of many algal species release substances that attract motile male (+ve) gametes. Among the seaweeds, *Ectocarpus* and *Fucus* have been extensively studied. The most recent work is that of Müller and his coworkers (Müller and Jaenicke, 1973; Müller et al., 1971).

To analyze the male attractant of *Fucus serratus* Müller and Jaenicke collected 252 kg of fresh female receptacles and from these obtained 9.8 liters of eggs and oogonia. From these eggs 690 μg of attractant were obtained. By gas chromatography and mass spectrometry the identity of the substance was determined as all-*trans*-1,3,5-octatriene(C_8H_{12}). This is a conjugated hydrocarbon. The substance was also produced synthetically.

The attractant in *Ectocarpus* was found to be S-1-but-1′-enyl-2,5-cycloheptadiene (Müller et al., 1971). Male gametes are strongly attracted to a female releasing this substance. The males become attached to the female with the tips of their front flagella. One male gamete fuses with the female and the remaining cells disperse. Later work by Müller (1976b) showed that the male attractant works on gametes that are sexually isolated. The final stage of cell fusion is blocked, but males continue to be attracted to female gametes.

The hormones of clumping behavior have not been studied extensively in seaweeds. Köhler (1956) has reported a clumping substance from *Chaetomorpha aerea*.

Environmental Effects

Environmental control of reproduction was reviewed extensively by Dring (1974). Of the environmental factors considered (light, temperature, chemicals, and exposure to air), light effects have been studied in most detail. Minimum light levels required for reproduction have been measured in several seaweed species. Presumably the observed response to light can often be interpreted as a photosynthetic response. Photosynthetic production of carbohydrates is required for the formation of reproductive bodies. This does not explain Müller's (1962) observation that in high light energies unilocular sporangium formation is favored in *Ectocarpus*, whereas plurilocular sporangia are favored in low light energies.

The effects of light color and light period have already been considered (see Chapter 4, "Responses of Plants to Lights"). Observations have also been made on the reproductive response of several seaweed species to various levels of temperature, nutrients, and exposure to air (Dring, 1974). At present we have no biochemical or physiological explanations for these responses. This contrasts markedly with our understanding of, for example, the role of phytochrome in the response of plants to photoperiod.

Endogenous rhythms

Endogenous rhythms of reproduction are those that can be shown experimentally or statistically to be to some extent independent of environmental changes (Sweeney and Hastings, 1962). The best docu-

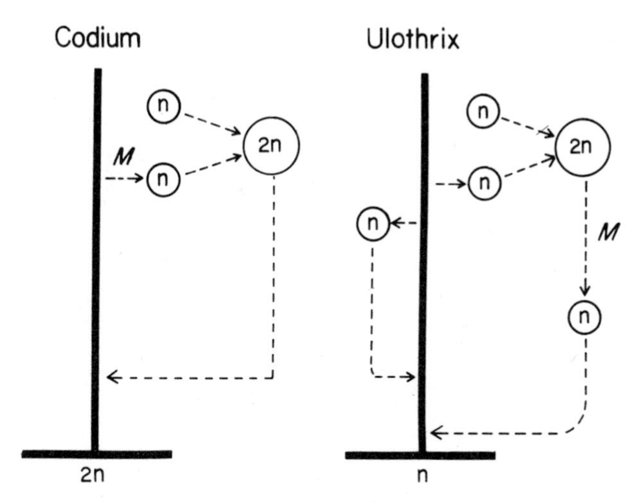

Figure 48. Life histories of *Codium* and *Ulothrix*, both of which have a single morphological phase (2n in *Codium*, n in *Ulothrix*). The vertical bars indicate morphological phases. Bodies in circles are gametes, spores, or zygotes. *M*, meiosis.

mented case of endogenous rhythmicity is the lunar rhythm of egg production in the brown alga *Dictyota*. Maturation of oogonia occurs every 2 or 4 weeks, corresponding to tidal cycles. Similarly *Ulva fasciata* in culture under constant conditions produces swarmers at the time of spring tides (Subbaramaiah, 1970). Page and Sweeney (1968) have demonstrated that a 4- to 5-day periodicity of gamete formation in the green alga *Derbesia* is controlled by an endogenous rhythm.

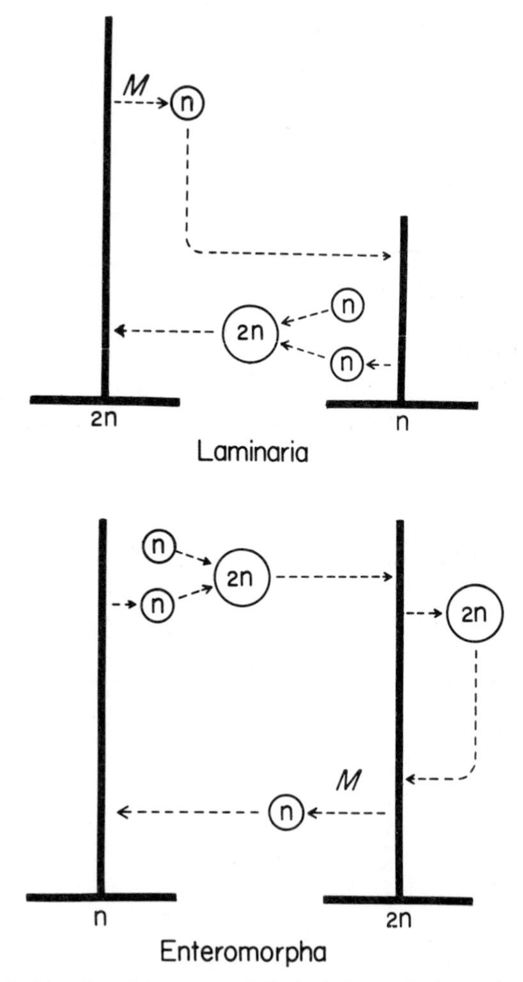

Figure 49. Life histories with two morphological phases. In *Laminaria* the two phases are morphologically dissimilar. In *Enteromorpha* the two phases are similar. *M*, meiosis.

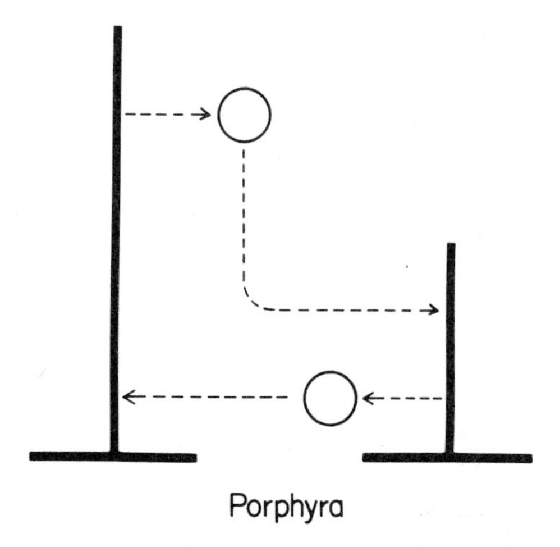

Figure 50. Life history of *Porphyra* according to Krishnamurthy (1959). The larger vertical bar represents the leafy intertidal thallus. The smaller bar represents the microscopic shell boring phase. Both phases have the same ploidy and there is no indication of meiosis or syngamy.

LIFE HISTORIES

The life history of an alga is the recurring sequence of morphological and cytological phases of the species under consideration (Drew, 1955). A morphological phase is the state of an organism recognizable by constant characteristic morphological appearance irrespective of chromosome number. A morphological phase begins with a single cell (spore or zygote) and ends with a single cell, the reproductive body it produces (Dixon, 1973). A cytological phase is the state of an organism characterized by mitotic divisions, all showing the same chromosome number. A cytological phase normally may be diploid or haploid. The diploid condition arises when fusion of two haploid nuclei occurs. The haploid condition arises by meiosis of a diploid nucleus.

In the seaweeds there may be one or more morphological phases in a given life history. Each of these gives rise to the next phase by the production of a single celled spore or zygote. The sequence from phase to phase is not necessarily obligatory, and a single phase may repeat itself through asexual or vegetative reproduction.

Tetraploid and triploid phases occur in *Ectocarpus siliculosus* (Müller, 1967) and *Plumaria elegans* (Drew, 1939), respectively. When there is one morphological phase in the life history it may be either

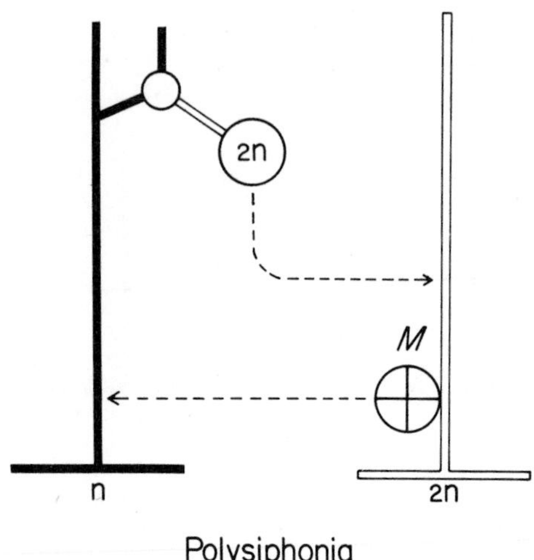

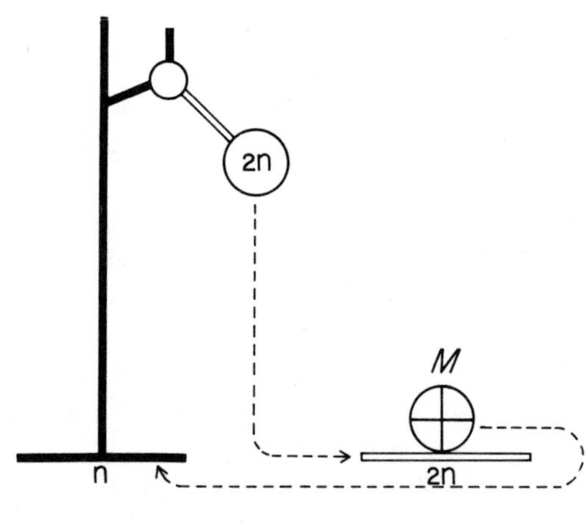

Figure 51. Life histories with three morphological phases. *Polysiphonia*-type showing haploid gametangial phase bearing the diploid carposporangial phase. The diploid tetrasporangial plant is morphologically similar to the gametangial phase. *Bonnemaisonia*-type where the gametangial and tetrasporangial phases are morphologically dissimilar. After Dixon (1973).

diploid or haploid. Examples exhibiting a sexual cycle are shown in Figure 48.

In some species there is no sexual cycle. For instance *Laminariocolax tomentosoides*, a brown alga, has a single morphological phase (apparently diploid) which repeats itself by the production of diploid spores (Russell, 1964).

Sexual life histories with two morphological phases are very common in the seaweeds. These two phases may be morphologically similar, as in *Enteromorpha*, or quite dissimilar, as in *Laminaria* (Figure 49).

According to Krishnamurthy (1959) and Richardson and Dixon (1968) some Bangiophyceae have two morphological phases, but these are of the same ploidy with no evidence of either meiosis or syngamy. (Figure 50). In the vast majority of the Florideophycean red algae there are three morphological phases in the life history. These are named on the basis of the reproductive organs that they bear (Dixon, 1973). Thus the gametangial phase bears gametangia, the carposporangial bears carposporangia, and the tetrasporangial bears tetrasporangia. The gametangial and tetrasporangial phases may be morphologically identical (*Polysiphonia*) or dissimilar (*Bonnemaisonia*). In all cases the carposporangial phase is born on the gametangial and is morphologically distinct from the other phases (Figure 51).

From this brief résumé we can see that in the seaweeds there are from one to three morphological phases and one to three cytological phases. The exceptions are innumerable and interesting. For instance, in *Palmaria palmata* only male gametangial and tetrasporic plants have been formed. Females seem to be absent. How then is the life history completed? Müller (1967) has shown that in *Ectocarpus siliculosus* gametangial plants may be diploid or haploid, whereas morphologically distinct sporangial plants are haploid, diploid, or tetraploid! Considerations of these phenomena are outside the scope of this book, but interested readers are referred to Dixon's (1973) treatment of the red algae and Wynne and Loiseaux's (1976) treatment of brown algae. The green algae, which are so interesting in this regard, have not been extensively reviewed recently.

Part III
Organization in Populations

Populations are assemblages of organisms belonging to a single species. The characteristics of seaweed populations considered here are demographic and genetic.

Chapter 6
Population Biology

DEMOGRAPHICS

Populations parameters are of three main kinds:

a. Size or density, which are affected by births, deaths, immigration, and emigration
b. Age distribution
c. Distribution of individuals in space

Very little information on these characteristics is available for seaweed species. Some examples of (a) and (b) are considered here.

In formal population studies schedules of mortality and reproduction are constructed and combined to give an "innate capacity for increase" in a given environment. No such complete analysis is available for any seaweed species. The nearest approach to a formal treatment is that of Rosenthal, Clarke, and Dayton (1974), who carried out a 5.7-year study of recruitment and survivorship in *Macrocystis pyrifera*, a giant kelp species. Rosenthal et al. (1974) used survivorship curves to study the mortality of plants (Figure 52). In September of 1969, 387 newly recruited *Macrocystis* sporophytes were counted in permanent quadrates. In the first month 231 of these died and by June, 1970, there were only five survivors. The tremendous loss of young plants means that the survivorship curve is concave. A curve of this shape is said to be type III, according to Pearl (1927). Organisms with type III curves are characterized by high reproductive rates and high juvenile mortality. The potential for reproduction is certainly enormous in kelps. Kain (1975) showed that enough spores are produced in a *Laminaria hyperborea* forest to cover the rock surface with 3.3 million spores mm^{-2} or, to put it another way, to form a layer 70 spores deep.

The data that are required for the construction of a survivorship curve form one column of a life table, which is a schedule of mortality for a population. The best way to construct a life table is by following a cohort of organisms through time. A cohort is a group of organisms recruited at the same time. A cohort life table for *Macrocystis pyrifera* (Table 7) has been constructed using survivorship data extracted from a published curve by Rosenthal et al. (1974). The most interesting column is e_x (average number of months of life remaining to those plants at the beginning of $month_x$). The shortest life expectancies are for the youngest age class and the oldest. All intermediate month classes have much larger life expectancies. Under certain conditions a life table may be constructed from knowing the age structure of a population at a given time. One of these conditions is a static age structure through time. Since a static age structure is most unlikely in disaster-prone seaweed populations, static life table construction is inadvisable. Nevertheless the age structure of a population may be very informative. Some seaweeds may be aged. *Laminaria hyperborea* is one of these. Stipes of this species have annual growth rings in much the same way as trees (Kain, 1971). Another ageable species is the fucoid *Ascophyllum nodosum*. In this plant one

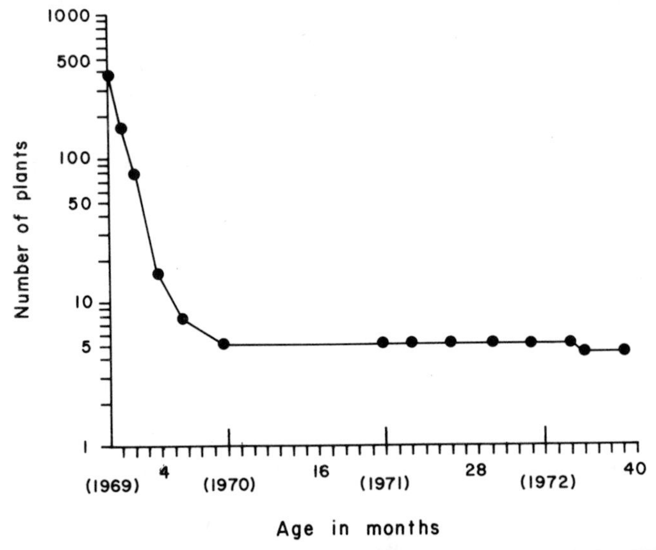

Figure 52. Survivorship curve of a cohort of *Macrocystis* sporophytes recruited in September, 1969. After Rosenthal et al. (1974).

Table 7. Cohort life table for *Macrocysitis pyrifera*

x^a	$l_x{}^b$	$d_x{}^c$	$1{,}000q_x{}^d$	$L_x{}^e$	$e_x{}^f$	$T_x{}^g$
0	387	371	958.65	201.5	0.65	252.5
4	16	11	687.5	10.5	3.18	51
8	5	0	0	5	8.1	40.5
12	5	0	0	5	7.1	35.5
16	5	0	0	5	6.1	30.5
20	5	0	0	5	5.1	25.5
24	5	0	0	5	4.1	20.5
28	5	0	0	5	3.1	15.5
32	5	0	0	4.5	2.1	10.5
36	4	1	250	4	1.5	6
40	4	0	0	2	0.5	2

Constructed from data in Rosenthal et al. (1974).

[a] x = age class in months.

[b] l_x = number alive at beginning of month$_x$.

[c] d_x = number dying during month$_x$, $d_x = l_x - l_{x+1}$.

[d] $1{,}000q_x$ = number dying per 1,000 alive at beginning of month$_x$. $1{,}000q_x = 1{,}000 \cdot d_x/l_x$.

[e] L_x = number of plants alive between month$_x$ and month$_{x+1}$. $L_x = l_x + l_{x+1}/2$.

[f] e_x = average number of months of life remaining at beginning of month$_x$. $e_x = T_x/l_x$.

[g] T_x = sum of life remaining to those aged$_x$. $T_x = \sum L_x$ from bottom up to and including month$_x$.

air bladder is formed on each shoot per year (no bladder is formed in year 1). Ages of shoots can be determined by counting bladders. The age profiles for *L. hyperborea* under very exposed and very sheltered stable conditions are shown in Figure 53. In the exposed population the larger plants are removed rapidly by wave action and this allows the development of smaller plants, hence the preponderance of individuals in age class 1. In contrast, the established forest canopy in the sheltered population inhibits the development of young plants, presumably through shading. It seems likely that the reproductive potentials of the two populations are quite similar, and that differences in age structure result from differences in intraspecific competition. In this respect the generalizations regarding age structure for animal populations may not be applicable.

GENETICS OF SEAWEED POPULATIONS

Interbreeding individuals in a single species population contribute to a common gene pool. Haploid or diploid organisms possess one or two

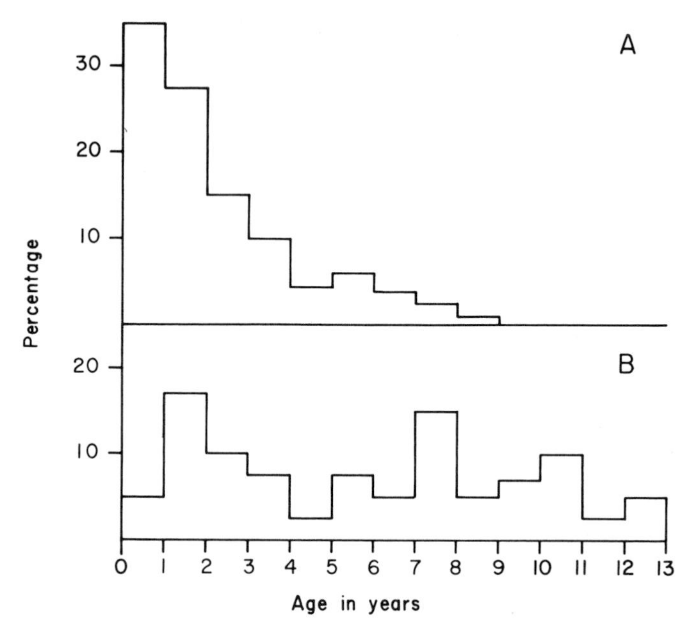

Figure 53. Age class structure of two populations of *Laminaria hyperborea*. Population A is exposed to strong surf action, population B is from sheltered, stable conditions. Redrawn from Kain (1971).

alleles at each genetic locus, respectively. However, in the common gene pool there may be tens of alleles for each locus. This illustrates one way in which populations have genetic characteristics that do not reside in the individual organisms. The discipline of population genetics is concerned with the control of gene frequencies in populations. Much of the theory in this field of study developed in the absence of a real knowledge of gene frequencies in wild populations. This situation was resolved in the early 1960s when electrophoretic techniques were developed that allowed the discrimination of enzyme forms produced by different alleles of the same locus. Thus without elaborate crossing experiments it was possible to determine allelic frequencies. Subsequently there has been a flood of publications and a significant advance in the field of populations genetics. Unfortunately seaweeds have mucilaginous carbohydrates, which interefere with the extraction of enzymes for genetic study. Consequently applications to these organisms have been few.

An interesting but limited study of allelic frequencies in *Enteromorpha* populations has been carried out (Hall, 1975). Two enzymes were studied, xanthine dehydrogenase and malic dehydrogenase. Each of these was found to occur in three alleloenzyme forms which could be separated electrophoretically. Six populations were sampled and the frequencies of the alleles determined. In three populations only a single genotype was found. In two of the populations several alleloenzymes were demonstrated (Table 8). At Melville Cove, Halifax, Nova Scotia, a considerable amount of genetic variation was found. One theory put forward by Hall (1975) to explain this variability is a founding effect—the plants establishing the population by chance possessed a variety of allelic combinations. Asexual reproduction of these plants would lead to cloning and the mixture of founding genotypes would be maintained without any crossing over at meiosis. Selection would determine the relative rareness of various alleles.

Cheney and Babbel (1975) have examined the degree of genetic relatedness in populations of several species of the red alga *Eucheuma*. Similarity was determined as number of shared alleloenzymes/total number of alleloenzymes. Generally, genetic similarity is related to morphological similarity. As in Hall's (1975) study, most of the loci occurred in only a single allelic form.

Information on genetic differentiation can be obtained experimentally by observations of phenotypes. A good example of this approach is the study of Russell and Bolton (1975). *Ectocarpus*, a filamentous brown alga, occurs in a variety of habitats of varying salinities. Plants were collected from three differing salinity regimes and

Table 8. Frequencies[a] of three alleloenzymes for two genetic loci for *Enteromorpha* in six populations from Nova Scotia[b]. Alleloenzymes for each locus were found to move at differing rates in an electrophoretic gradient

Locality	Number of plants analyzed	Numbers found of each alleloenzyme					
		MDH_S	MDH_I	MDH_F	XDH_S	XDH_I	XDH_F
Melville Cove	90	12	6	70	10	7	73
N.S. Yacht Squadran Basin	87			35	2	1	84
Peggy's Cove	86			25			86
Paddy Head	30			15			30
Shag Bay	21			20			21
Wynacht Point	140	2	25	113	25		109

After Hall (1975).

[a] Unsuccessful electrophoretic visualization results in unbalanced numbers.

[b] Fast, slow, and intermediate forms of xanthine dehydrogenase (XDH) and malate dehydrogenase (MDH) were identified.

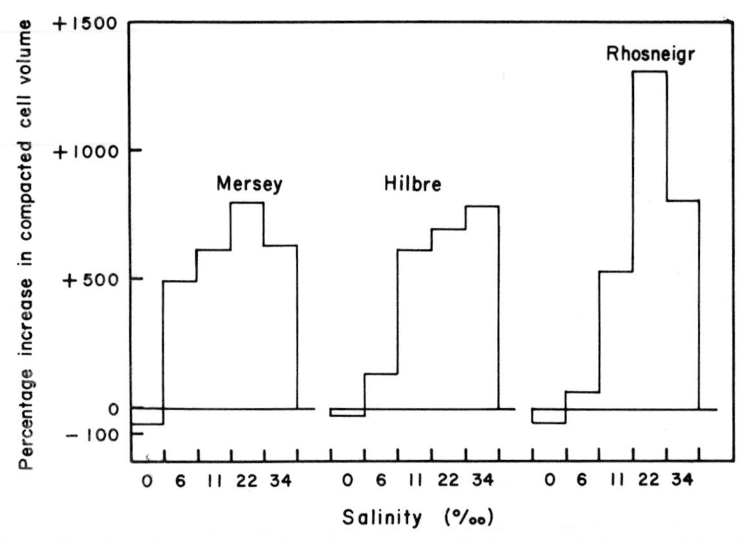

Figure 54. Growth of three strains of *Ectocarpus siliculosus* in relation to salinity. The Rhosneigr strain originated from high salinity conditions, the Mersey strain from dilute estuarine water, the Hilbre strain from intermediate conditions. From Russell and Bolton (1975), *Estuarine and Coastal Marine Science*, Vol. 6, p. 93. Copyright by Academic Press Inc. (London) Ltd.

maintained in culture in full seawater for a number of years. The strains were then treated with a range of salinities from 0 °/oo to 34 °/oo and their growth was followed (Figure 54). A better growth performance at low salinities was demonstrated by strains originating from low salinity habitats. The results indicate that a degree of genetic differentiation occurs between populations.

COMMENT

The shortness of this chapter is a reflection of the minimal attention that has been given to the population biology of seaweeds. This would seem to be a most promising area both academically and commercially. A knowledge of rates of recruitment and mortality integrated in formal schedules would provide an excellent basis for the management of commercial species when coupled with a knowledge of growth rates. The existence of unusual life histories provides interesting material for genetic analysis of populations.

Part IV
Organization in Communities

According to Mills (1969) a community is a group of organisms occurring in a particular environment, presumably interacting with each other and with the environment and separable by ecological survey from other groups. Chapter 7 deals with the structural organization of the population assemblages on the seashore. Chapter 8, goes on to examine the mutually regulative and interdependent aspects of this structure.

Chapter 7
Community Structure

The structuring in space of population assemblages lends itself to many different forms of analysis. Interest may be directed to the vertical structure, since communities usually show vertical stratification of growth forms. Horizontal differentiation is also found, and here we are concerned with the patterns of distribution of species assemblages. In seaweed communities interest in horizontal differentiation has revolved almost exclusively around the classification of communities. Many different methods of classification of plant communities have evolved, and several of these have been applied to the seashore vegetation:

1. Classification by dominant species: A measure of dominance for seaweed vegetation is commonly arrived at in a subjective way. In cold-temperate seas the greatest biomass proportion in any given community usually belongs to a single species, and this is regarded as the dominant. Thus the vegetational units are defined by a series of dominance types.
2. Numerical classification: The degree of similarity between quadrate samples is used to construct a hierarchical classification. Alternatively, the relative distributional similarity of species may be used.
3. Florisitic classification: All of the plant species are considered together to produce a taxonomic hierarchy, the basic unit of which is the association. This is the phytosociological approach.

VERTICAL STRUCTURE

Cursory examination of a seashore at low tide shows a mat of seaweeds lying in heaps on the rocks. Very little impression of vertical structure is obtained. However, the advent of scuba and its application to marine biology has revealed the vertical profiles of some of the well-developed seaweed forests in the cooler oceans. An excellent example is seen in the work of Dayton (1975a), who examined the kelp beds of Amchitka Island, Alaska. His profile diagram is shown in Figure 55. Four layers were detected in the vertical structure in deep water. *Alaria fistulosa* is a large kelp that may grow up to 20 m in length. The fronds float and form an uppermost canopy layer lying on the surface. Below this canopy there is a layer of stipitate *Laminaria* species. The lowermost kelp layer lying below the Laminarias is made up of *Agarum cribosum* which lies in flat sheets on the substratum covering the final layer of vegetation, a turf of small red and green algae.

A novel approach to studying vertical structure of seaweed vegetation has been developed by Neushul (1972). He made use of side scan sonar, which produces a dynamic picture of the movement of the layers of vegetation in relation to water movement.

den Hartog (1959) has recognized the vertical structuring of intertidal seaweed vegetation. On the shores of the North Sea he dis-

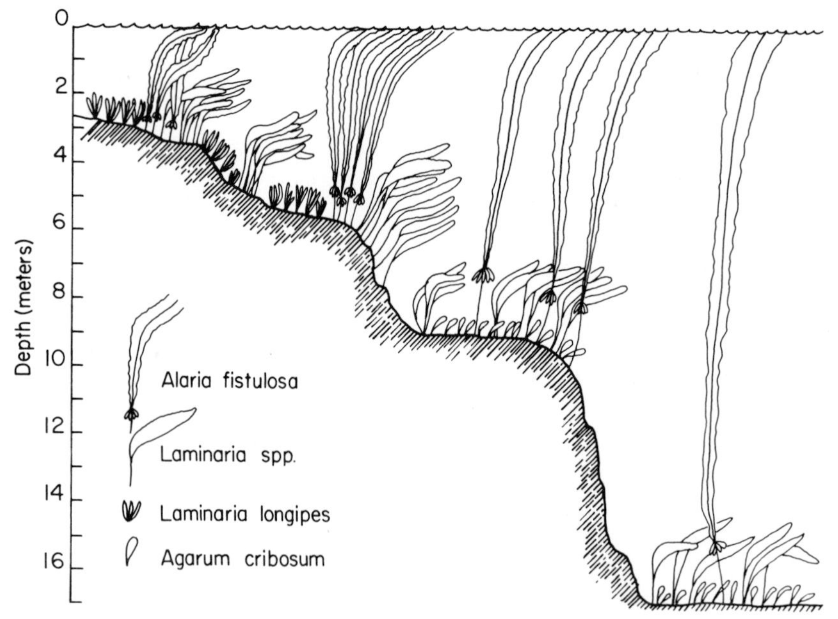

Figure 55. Profile diagram of the kelp community at Amchitka Island, Alaska. After Dayton (1975a).

tinguished three layers. A primary stratum consists of encrusting algae flattened to the substratum. Above this is a secondary stratum of small foliose forms and a tertiary stratum of the large brown alga *Fucus*.

CLASSIFICATION INTO ZONES OF DOMINANTS

Dominant species may be determined in one of two ways. The first emphasizes a descriptive definition of importance often, in the case of seaweeds, based on relative abundances or biomass. The second approach, developed by Paine (1974) and Dayton (1971, 1975b), emphasizes a functional definition of dominant species based on the relative impact these species can exert on other species in the community. The descriptive approach is taken here and examined in relation to the structure of seaweed communities.

On a rocky shore in cool temperate regions of the world even a casual observer is impressed by the fact that the organisms are arranged in horizontal strips or zones (Figure 56). Quite often these organisms differ in color and the transition between zones is sometimes so sharp it is as though a pencil line had been drawn between them. So impressive is this phenomenon that its description became a major occupation of seaweed ecologists. This description of zones has been undertaken on a worldwide basis by the Stephensons, who spent many years touring the world and, as they put it, "reading" the shores. Their work over a 40-year period culminated in the publication of a book, *Life Between Tidemarks on Rocky Shores* (Stephenson and Stephenson, 1972). Other works have not been on such a geographical scale. Lewis (1964), for instance, gives a profusely illustrated guide to the shores of the British Isles, and there have been innumerable other works that have listed the vertical distribution of algal species on this or that shore. The Stephensons (Stephenson and Stephenson, 1949) attempted to produce the first real unifying theme to all of these studies—a basic plan of universal features of zonation. This has been modified into the generally accepted scheme of Lewis (1961), which is shown in Figure 57. According to this concept the horizontal belts shown can be found on any rocky shore. The dominant marker species may vary, but the three zones are held to be universal. The dominance of the species used as markers was not measured in any quantitative way by the architects of the scheme. Nevertheless this dominance probably does have some basis in the overwhelming biomass of the dominants over the dominated species.

Figure 56. Intertidal zonation on a vertical rock face in the Tsitsikama Coastal Marine Park, South Africa. The upper surface is barren except for a few specimens of orange lichens. Below this is a black band of blue-green algae which is followed by a zone of barnacles. The barnacle zone gives way to a zone of pink encrusting coralline algae with the distinctive limpet *Patella cochlear*. The lowermost zone above the water consists of the foliose coralline red algae, mainly *Amphiroa ephedraea*.

The universal scheme of zonation is the only major unifying concept to emerge from over 75 years of classification of seaweed communities into zones of dominant species.

NUMERICAL CLASSIFICATION

Russell (1972) points out that the recognition of zones by dominant species seems afflicted by a circular argument. Thus three sorts of

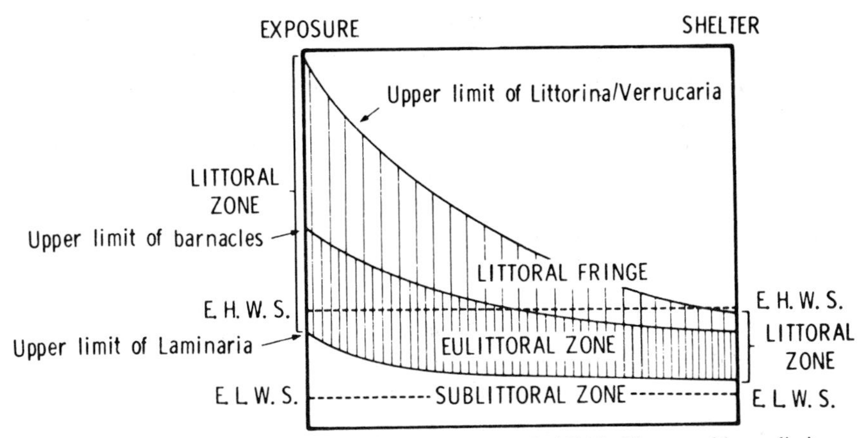

Figure 57. Three universal zones recognized by Lewis (1961). Upper and lower limits, marker species, and tide levels are indicated. E.H.W.S., extreme high water spring tide; E.L.W.S., extreme low water spring tide.

organisms with different and more or less contiguous vertical distributions are chosen to define three zones, which are then identified by the presence of the organisms. One way around this problem lies in the use of numerical techniques. The numerical analyses of Russell (1972, 1973) and of Prentice and Kain (1976) are discussed here.

Russell carried out his first analysis on a shore in which the sublittoral zone was absent. The eulittoral and littoral zones were evident. Quadrates of a predetermined suitable size were placed at random on the shore (two were placed subjectively). Frequency and cover values of species were recorded for each quadrate. Using these kinds of data, two types of analysis were carried out, cluster analysis and analysis of community similarity.

The first step in a cluster analysis is to compute the degree of association between pairs of species. Russell used the straightforward technique of χ^2 analysis, which is explained as follows: There are n quadrates, a containing species A and B, b containing species B alone, c containing species A alone, and d containing neither. A normal 2 × 2 contingency table is made up:

		Species *A* +	Species *A* −	
Species *B*	+	a	b	$a + b$
	−	c	d	$c + d$
		$a + c$	$b + d$	

$$\chi^2 = (ad - bc)^2 \cdot \eta / (a + b)(c + d)(a + c)(b + d)$$

χ^2 is an index of association between species. An index of association is usually represented by the symbol I, and values between pairs of species are now entered in association matrix. All values of I are made positive.

	Species			
	A	B	C	D
A	–	I_{AB}	I_{AC}	I_{AD}
B	I_{AB}	–	I_{BC}	I_{BD}
C	I_{AC}	I_{CB}	–	I_{CD}
D	I_{AD}	I_{DB}	I_{DC}	–

The values of I are summed for each species across the rows. The species with the highest $\sum I$ is called the critical species and the data are then split into two groups, those with the critical species and those without. The analysis is then repeated with the two subgroups, which are then split again, and so on. The divisions take place at progressively lower levels of $\sum I$, thus we can easily construct a dendogram, as shown in Figure 58.

Russell's other approach was analysis of community similarity. In this case Russell used 21 quadrates and computed Czekanowski's coefficient of similarity between each pair:

$$CC = (2c/a + b)100$$

where a = number of species in quadrate A, b = number of species in

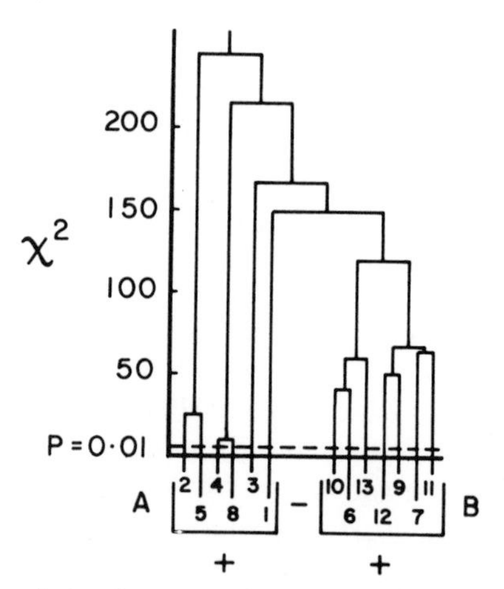

Figure 58. Cluster analysis of common intertidal species using χ^2 as an index of association. Species are 1, *Prasiola stipitata;* 2, *Blidingia minima;* 3, *Lyngbya semiplena;* 4, *L. lutea;* 5, *Fucus spiralis;* 6, *Elminius modestus;* 7, *Enteromorpha prolifera;* 8, *Schizonema* sp.; 9, *Fucus vesiculosus;* 10, *Ulva lactuca;* 11, *Cladophora rupestris;* 12, *C. sericea;* 13, *Fucus serratus*. After Russell (1972). By permission of Blackwell Scientific Publications Ltd.

quadrate B, and c = number of species common to both. Russell weighted his calculations of the coefficients for frequency and then for cover. The end result is a matrix of between-quadrate similarity coefficients. These he arranged by hand sorting so that similar quadrates occurred close together in the matrix (Figure 59).

Notice that in both analyses the data fall into two groups. These correspond nicely with the previously recognized littoral fringe in the upper shore and the midshore eulittoral zone.

In 1973 Russell extended his work to a three-zone shore where the sublittoral zone as defined by laminarian species was present. He carried out similar analyses, extending the work by using two additional indices of community or vegetation similarity. The distinction between the eulittoral and littoral fringe was again marked. This discontinuity is known as the "litus" line, a term introduced by Sjöstedt (1928), who regarded it as the ecological boundary between land and sea. However, the discontinuity between the sublittoral and eulittoral zones was by no means distinct in Russell's analysis.

Prentice and Kain (1976) have carried out a similar analysis of sublittoral vegetation. Three subjectively defined zones were recognized. Immediately below low water there was a zone dominated by the large kelp *Laminaria hyperborea*. This was followed in deeper water by a zone dominated by another kelp, *Saccorhiza polyschides*. Below this zone was an open community distinguished by large numbers of the sea urchin *Echinus esculentus*. These three zones did show up in a numerical analysis, but not with the distinctness of the littoral fringe/eulittoral discontinuity.

What are the general conclusions to be drawn from these numerical analyses? It seems that the definition of communities by dominant species as in the Stephenson school does not produce a scheme that is clearly representative of the total vegetation on the seashore. Only the littoral fringe appears to be a clearly distinct vegetational unit. Even then care must be taken in interpretation. The monothetic divisive classification used by Russell is designed to split the data up into units. Because it is possible to classify a collection of quadrates, it does not follow that the vegetation that they represent is classifiable into well-defined separate parts.

FLORISTIC CLASSIFICATION

Floristic classification is the objective of phytosociologists. The grand design of this group of botanists is to produce a taxonomic hierarchy of vegetational units in the same way that Linnean species are

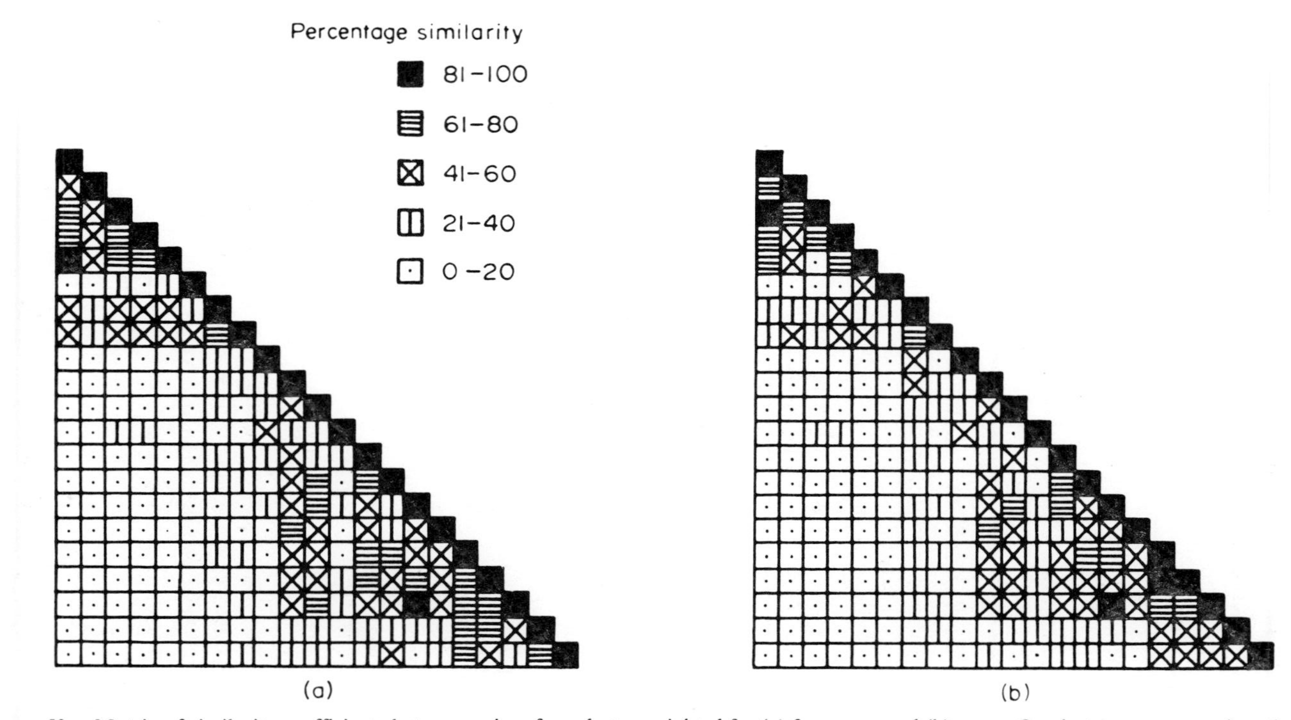

Figure 59. Matrix of similarity coefficients between pairs of quadrates weighted for (a) frequency and (b) cover. Quadrates are arranged on the vertical and horizontal axes, (hence 100% similarity on the diagonal) and have been shuffled so that similar quadrates are situated adjacent to one another. From Russell (1972). By permission of Blackwell Scientific Publications Ltd.

agglomerated into genera, genera into families, families into orders, and so on. There are many difficulties in the application of Linnean taxonomy, but no one doubts that there are hierarchical relationships between taxa. This is not so in the case of phytosociological classification.

Phytosociology has not developed to any great extent in the English speaking world and neither has its application to seaweed vegetation. Most marine work has been carried out in the Mediterranean. Two main schools of phytosociology have developed in Europe, the Upsalla school and the Zurich-Montpellier group of Braun-Blanquet. The aims of the two schools differ little, but the methods are quite dissimilar. The Zurich-Montpellier approach is much more subjective and dependent on the experience of the investigator. This second approach has been used by marine phytosociologists. Sample plots are selected on the basis of a visual assessment of vegetational homogeneity. The importance of species in the sample plot is determined by two coefficients, one of cover-abundance, and the other of sociability. The cover-abundance index of each species in the plot is recorded. This index is on a 6- or 7-point scale:

5 covering more than ¾ of sample plot
4 covering between ½ and ¾ of sample plot
3 covering between ¼ and ½ of sample plot
2 with any number of individuals covering $\frac{1}{20}$ to ¼ of sample plot, or for very numerous individuals covering less than $\frac{1}{20}$ of the area
1 numerous, but covering less than $\frac{1}{20}$ of the sample plot, or fairly sparse, but with greater cover value
+ sparse and covering only a little of the sample plot
r rare and covering only a small part of the plot, usually only one specimen in plot

Sociability is estimated on a 5-point scale:

5 in large, solid stands, very dense populations
4 in small colonies or mats, rather dense populations
3 in small patches, distinct groups
2 in small groups or clusters or tufts
1 growing singly

Thus the raw data consist of a species list, each species with a coefficient of abundance-dominance and a coefficient of sociability. Lists from different community samples are compared in a table in which the rows are species and the columns community samples. A list from den Hartog (1959) giving only cover-abundance ratings is shown in Table 9. By visual examination, many tables are compared and the basic taxonomic unit or "association" abstracted. The association is a community with more or less constant floristic composition as well as important character species. Character species are present in the samples of one community type, but absent or less important in others.

The association has rank equivalent to that of the species in Linnean taxonomy. Below it there are subassociations, variants, and facies. Above it there are alliances, orders, and classes.

The phytosociology of the Mediterranean is reviewed briefly by Edwards et al. (1975). Most workers follow Molinier's delimitation of communities on rocky shores. There are said to be three communities, that of the supralittoral zone (littoral fringe) and those of the upper and lower mediolittoral (both included in the eulittoral zone). The lower mediolittoral is divided into several facies. The sublittoral zone consists of two communities, the photophilic and the coralligenous. The latter community occurs in regions of feeble illumination and the former in well-lighted areas. Boudouresque (1970, 1971) believes that distinct communities cannot be recognized and that the vegetation consists of a complex continuum of populations.

Apart from the Mediterranean, marine phytosociology has developed in Japan (Saito and Atobe, 1970; Saito et al. 1971; Taniguti, 1962) and in the Netherlands (den Hartog, 1959). These areas are strongly influenced by tides and seaweed zonation is

Table 9. Association table for part of the upper shore on the coast of the Netherlands[a]

Sample plots	1	2	3	4
Percentage of covering	100	55	95	85
Verrucaria maura	5	4	5	4
Lichina confinis	–	–	+	2
Hildenbrandtia prototypus	1	–	–	–
Verrucaria mucosa	+	–	–	–
Rhodochorton purpureum	+	–	–	–
Pelvetia canaliculata	–	–	+	–
Caloplaca marina	+	–	+	2
Lecania erysibe var. *erysibe*	+	–	–	–
Xanthoria parietina	–	–	–	1
Caloplaca murorum	–	–	–	1
Ramalina siliquosa	–	–	–	2

After den Hartog (1959).

[a] For each species a cover-abundance index is given.

Table 10. Associations found on the coast of the Netherlands

Littoral fringe	*Verrucaria maura* association *Calothrix scopulorum* association
Eulittoral	*Pelvetia* association *Fucus spiralis* association *Fucus vesiculosus* association *Ascophyllum nodosum* association *Fucus serratus* association *Polysiphonia-Chaetomorpha* association
Sublittoral	*Laminaria* association

After den Hartog (1959).

prominent. Thus the phytosociology tends to revolve around zonation. This is shown in den Hartog's (1959) list of associations found in the Netherlands (Table 10). Any phycologist with Stephensonian inclinations would immediately recognize the associations as zones.

The Dutch contribution to marine phytosociology is continuing through the work of Van den Hoek, Cortel-Breeman, and Wanders (1975). The methods used in the study are interesting in that the conventional Braun-Blaunquet approach was supplemented by a numerical analysis of the raw data tables. A cover-abundance index was assigned to each species. A cluster analysis of all of the relevés[1] was carried out after arc sine transformation of the Braun-Blaunquet abundance estimates to give importance values of each species. Similarity between pairs of relevés was calculated as follows: $S = \sum \min(x_i y_i)/\sum x_i + y_i - \sum \min(x_i y_i)$, where S = similarity; x_i, y_i = importance values of species i in relevés x and y. Cluster analysis was carried out by an agglomerative technique. Application of this method to the coral reefs of the Netherlands Antilles showed six zonal communities, each with its own assemblage of organisms, inhabiting six different depth zones. Once again there is an almost inevitable return to the concept of zonation of marine organisms.

CONCLUSIONS

Seaweed ecology had its origins in the recognition of zones based, quite subjectively, on certain marker or dominant species. Subsequent

[1] A relevé refers to a site analyzed with detailed records of total species present, their relative importance, and other analytic characters of the flora and habitat.

phytosociological analyses erected associations which, in the intertidal region, correspond to zones of dominants. Numerical analysis reveals only one sharp vegetational discontinuity. This falls at the transition of the littoral fringe and the eulittoral zone. Other zonal discontinuities do not delimit well-defined vegetational units. This certainly supports the contention of Boudouresque and Lück (1972) that marine vegetation consists of a continuum of species assemblages rather than distinct communities.

Chapter 8
Community Interactions

The distribution of seaweed species on the shore is effected by biotic and abiotic interactions. Whittaker (1962) is of the opinion that most species in plant communities do not interact. In other words, species are distributed independently along environmental gradients. The concept of the web of life, in which species populations are bound together, is rejected. Most of Whittaker's views are based on observations of forest systems. Forest trees are very long lived and do not lend themselves to experimental analysis of species interactions. Seaweed vegetation, on the other hand, furnishes excellent experimental material for analyzing community interactions. The feasibility of this approach was realized in the 1960s by Paine and has been extensively developed by his students.

Although the most significant experimental analyses of the distribution of organisms on seashores are now primarily focused on biotic interactions (competition, predation), the study of physiological tolerance limits has a much longer history. This aspect is treated next and is reviewed extensively by Zaneveld (1969).

ZONATION

Physiological Tolerance to Abiotic Effects

As mentioned in the previous chapter, the occurrence of zones is the most prominent aspect of the structuring of intertidal organisms (sessile plants and animals). The zoned species have rather sharp distributional limits. The overriding question for consideration is: what causes the appearance of a distributional hiatus? Intertidal organisms live in a tidal environment and there has been a search for an explanation in the rhythms of submergence and emergence (Doty, 1946). Others have rejected the tidal hypothesis (Stephenson and Stephenson, 1949) in favor of other physicochemical factors, such as spray or salinity gradients (den Hartog, 1968). Whatever their differences, this group of ecologists sees species occurring within and to the limits of their physiological abilities in the manner shown in Figure 60.

Sharp zonation of monocultures can be produced in tidal machines (Townsend and Lawson, 1972). However, on the shore species zones are commonly contiguous, so that the lower limit of one species marks the upper limit of another. Does this mean that the line of the discontinuity marks the physiological tolerance limits of the two species? This is really the "crunch" question and it has never been answered affirmatively by reliable experimentation. This is not to say that species on the shore do not show varying tolerance limits; rather, it has yet to be shown that zoned contiguous species have nonoverlapping tolerance ranges.

The occurrence of noncontiguous zonal limits is quite different. For instance the upper limit of the blue-green algal zone or of the

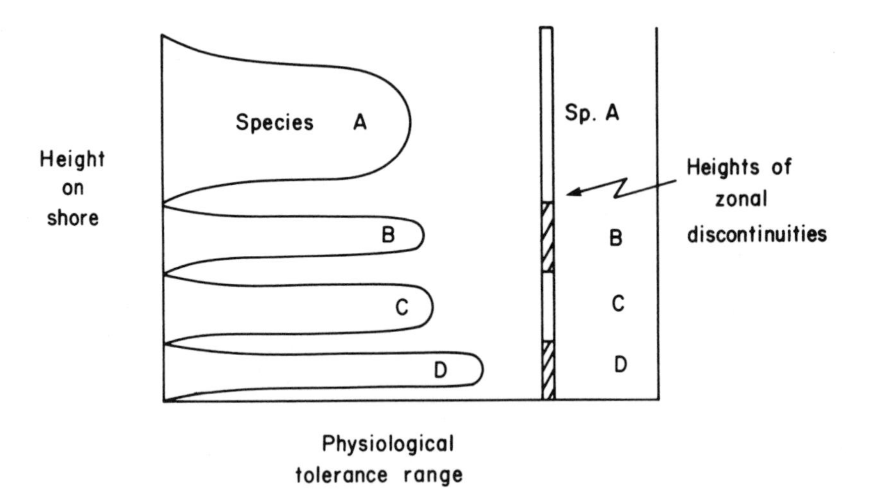

Figure 60. Vertical arrangement of four zoned species (A, B, C, and D) and their physiological tolerance ranges assuming that the extent of species zones corresponds to the extent of the tolerance ranges. Note that according to this explanation tolerance curves for contiguous species must begin and end coincidentally.

barnacle *Chthamalus* (Connell, 1961) occur on bare rock. The lack of interspecific interactions and the experimental demonstration of zonation in monocultures by Townsend and Lawson (1972) supports a physiological explanation in these cases.

Let us examine now some of the approaches that stress physicochemical effects on the distributional discontinuities on seashores. The best known of these is Doty's Critical Tide Factor Hypothesis (Doty 1946). The daily tidal regimen on the west coast of North America is shown in Figure 61. There are two high tides and two low tides in a single day. The two highs are of different heights, as are the two lows. Figure 61 shows that there is a gradual increase in emergence up to the 3.5-foot tidal mark, then suddenly at LLHW there is an increase in a single exposure time to 10.25 hr. From LLHW there is a gradual increase up to an emergence time of 23 hr at 5 feet. However, just above this level the emergence time is 144 hr. Thus we do see a smooth time-height curve of the form in Figure 62. The real situation is shown in Figure 63. Doty found that at the points of rapid change in submergence-emergence times zonal discontinuities of seaweeds occur. Unfortunately, as Connell (1972) has pointed out, relatively few of the upper and lower limits of algal species actually occurred at these sudden points of change.

Doty and Archer (1950) also designed an experimental test of the Tide Factor Hypothesis. Seaweeds were treated with 0.001 M formaldehyde, seawater concentrated 1.5 times, and a temperature of 27.5°C for varying durations of time. When the treatment time was doubled or tripled over that time at which only just significant injury took place, death of the alga usually occurred. This was taken as evidence in favor of the critical Tide Factor Hypothesis because, as is discussed above, at the critical tide levels there are two- or threefold changes in the emergence-submergence patterns. There is something of a conceptual flaw in this argument because changes in physicochemical parameters simulating those occurring at the critical tide levels were not tested. Put this way, the use of formaldehyde is seen to be irrelevant. The important question remains unanswered: at the zonal discontinuities (or critical tide levels) do conditions represent the extreme of the physiological range for the species under examination?

The importance of this question can be seen in the following example. *Pelvetia canaliculata* and *Fucus spiralis* are North Atlantic fucoids restricted to relatively high intertidal zones. Fischer (1929) found that both of these species will die if continually submerged. Is this evidence that death from submergence causes zonation of these

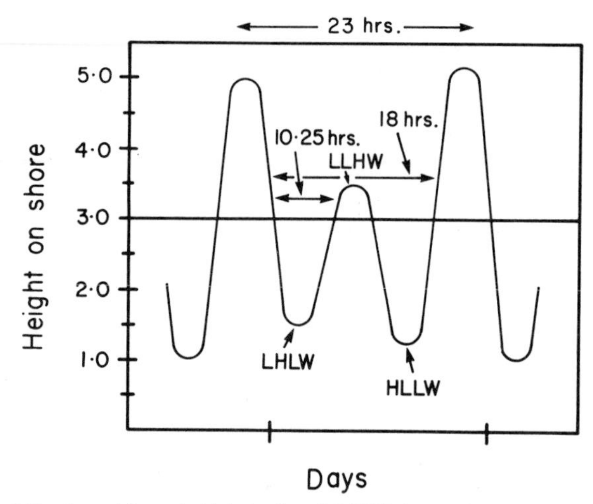

Figure 61. Mixed semidiurnal tidal rhythm. LLHW, lowest lower high water; LHLW, lowest higher low water; HLLW, highest lower low water. See text for details. After Doty (1957).

species? The answer is no, because the species have their lower limit above mean tide level, where they are submerged less than half the time. To a greater or lesser degree, flaws of this kind can be seen in virtually all of the published experimental work dealing with physico-chemical control of zonation. There is no doubt that species living at different levels on the shore respond differently to desiccation, high

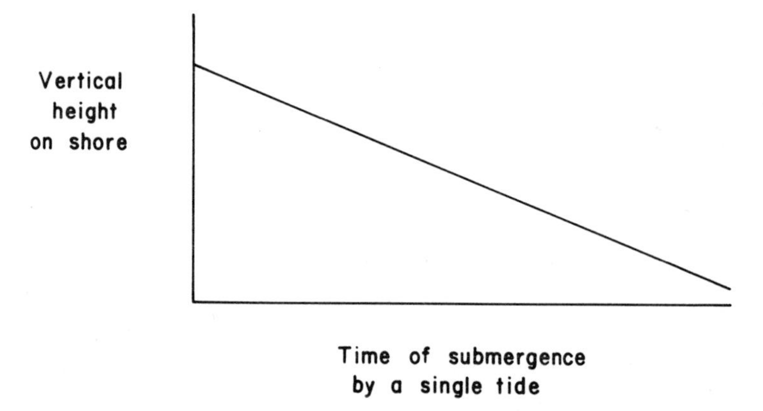

Figure 62. Tidal movements do *not* produce a smooth time-height gradient in the manner shown here. The gradient is broken into discontinuities, as shown in Figure 63.

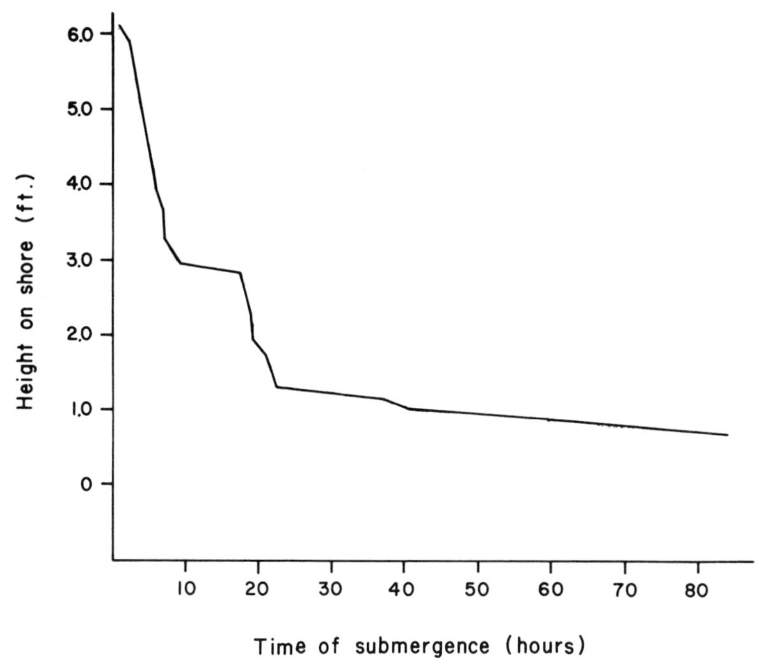

Figure 63. Time-height gradient produced by the action of mixed semi-diurnal tides on the Pacific coast of North America. Notice a series of discontinuities. After Doty (1946).

temperatures, light color, salinity, and so on. However, evidence of this kind is not sufficient to establish a physiological explanation for contiguous zones of distribution.

The Tide Factor Hypothesis is the best known explanation for the structuring of intertidal communities by abiotic factors. There have been detractors however. Stephenson and Stephenson (1972) stated, "We realize that the popular idea that zonation is caused by tides is itself untrue." As the Stephensons point out, zonation occurs above and below water level and around nontidal waters. These authors feel that zonation results from the existence of an interface between water and air.

den Hartog (1968) has developed a theory of zonation in which salinity plays a major role. During submersion intertidal organisms are in an environment of relatively constant salinity. During emersion desiccation may increase salinity levels, or rain may dilute the salt content. Just as Doty (1946) determined tidal discontinuities, den Hartog (1968) was able to discern a discontinuous increasing freshwater influence as one progresses through the sublittoral, eulittoral, littoral fringe, and maritime zones. Five discontinuous belts of fresh water influence were established:

	Tidal heights	Salinity characteristics
1.	Mean low water spring-mean low water neap	High, small fluctuation
2.	Mean low water neap-mean high water neap	Some fluctuation
3.	Mean high water neap-mean high water spring	Large fluctuations
4.	Above mean high water spring	Extreme fluctuation
5.	Sea spray reaches this level only seasonally and during storms	Low, but rarely increases

It must be stressed again that den Hartog (1968) proposed discontinuities in the freshwater influence between each of these levels. den Hartog also recognized that air and light act with salinity and are the three major abiotic factors dominating the intertidal zone. However, he concluded his treatise with the statement that the joint action of the three abiotic factors is insufficient to explain the often sharp demarcation between the belts of species. Biological competition is invoked as a fourth factor that plays a role in zone formation. The way in which biotic interaction may manifest itself in the distribution of organisms into zones was not examined.

den Hartog's (1968) realization of the role of biotic interaction contrasts with that of Doty (1957), who rejected what he termed the "trailside" (competition) scheme for zonation.

Biotic Interactions

The majority viewpoint among phycologists is that the zonation of seaweeds is primarily a result of the differential tolerance of species to the range of conditions on the shore. The range of conditions is seen by some as a gradient (Townsend and Lawson, 1972) and by others as a series of discontinuities (Doty, 1946). Biotic effects are admitted, although rather reluctantly, to have a modifying role. The thesis put forward here is that biotic interactions are the primary effects with abiotic modifications. This is also the view of Menge (1976), who states that "the mechanism maintaining patterns of community structure in New England (including richness) are primarily predation (all rocky shores but exposed headlands) and interspecific competition (exposed headlands) with important, but secondary influences of physical and biological disturbance."

All experimental evidence points to the fact that when two species populations meet at a sharp boundary neither of the species has reached the limits of its physiological tolerance range. On the other hand, when a species ends its distribution on bare rock we must conclude that it has reached the limit of its tolerance range. Figure 64 illustrates these concepts.

What is the evidence for these ideas? The complexity of the interactions between species is illustrated well by the work of Paine (1971). Working on the west coast of New Zealand, Paine observed that three organisms dominated the intertidal. The top carnivore is a starfish, *Stichaster australis*, which preys on a mussel, *Perna canaliculus*. *Perna* commonly forms a distinct zone in the mid-intertidal. The third dominant species is the bullkelp *Durvillea antarctica*, which forms a zone in the low intertidal. Paine used two experimental treatments. In the first, *Stichaster* was exluded. In the second, both *Stichaster* and *Durvillea* were removed from experimental areas. The results were striking. Within 8 months the mussel was able to colonize areas 60 cm vertically below control areas. This was a 40% increase in species range. In areas where *Stichaster* is

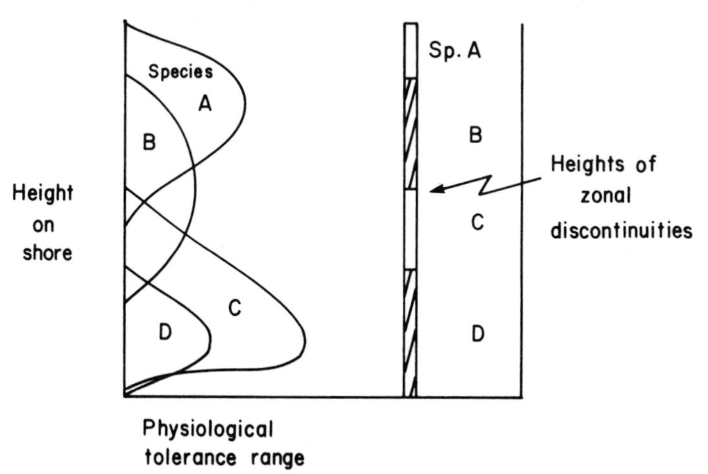

Figure 64. Vertical arrangement of four species (A, B, C, and D) where their observed ranges are of smaller dimensions than their physiological ranges. Note that the tolerance ranges overlap. The discontinuities are caused by biotic interactions. Compare with Figure 60.

overcollected in New Zealand the mussel *Perna* establishes a line of contact with the *Durvillea* zone. However, Paine observed that *Perna* was unable to establish itself under the *Durvillea* canopy. When both *Durvillea* and *Stichaster* were removed, *Perna* was able to extend its distribution to the lower limit of the *Durvillea* clearing.

These experiments show that the number of zones on the shore depends on: a) the predatory activities of *Stichaster* on *Perna*, and b) the competitive interaction between *Perna* and *Durvillea*. It is possible to change a 3–4 zone region in the mid- to low intertidal into a single zone by altering biotic relationships.

The types of relationships revealed by Paine (1971) have been repeated in other geographical areas (Dayton, 1975b; Menge, 1976; Paine, 1974). Although the work of Paine and his "Washington" school is rather recent, it must be pointed out that experimental evidence of this kind was available 25 years ago when the Critical Tide Factor Hypothesis was in its heyday. Burrows and Lodge reported in 1951 on the results of an experiment in Britain in which all herbivorous limpets (gastropod mollusks) and larger algae were removed from a 10-m wide strip extending through the intertidal. On a well-zoned shore in Britain contiguous zones are found of *Fucus spiralis* (high intertidal), *F. vesiculosus* (mid-intertidal), and *F. serratus* (low intertidal). On the experimentally treated strip all of the species were found throughout the intertidal zone. In other words all three species of *Fucus* are normally confined to zones that are of much smaller dimensions than the potential or pre-competitive niches.

The importance of biotic interaction in the formation of sharp boundaries between pairs of zoned species is incontestible. But this statement does not explain how, for instance, competition can effect observed distributions. To find an explanation we have to delve into the theories of population biology. When a population is growing continuously the rate of population growth is expressed as $dN/dt = rN$. dN/dt represents the change (d) in N, which is the number of organisms with a change in time (t).

r is a constant termed the instantaneous rate of increase. It has two components: b = the individual birth rate, d = the individual death rate. Thus, $r = b - d$.

Populations do not grow continuously because they exist in a resource-limited environment. Therefore we have to introduce a new term into the equation:

$$dN/dt = rN\,(K - N)/K$$

K is the carrying capacity of the environment, that is, the theoretical density a population could attain in a given environment. Since $K/K = 1$, the value of $K-N/K$ will approach 1 when N is very small. When N is close to K, $K-N/K$ becomes small and dN/dt approaches 0.

The equation of population growth can also be used to include two species. When two species populations N_1 and N_2 are simultaneously introduced into an environment, the rates of population increase will be r_1 and r_2, respectively. The carrying capacities will be K_1 and K_2.

The inhibiting effect of one individual on its own population growth is $1/K_1$; its inhibitory effect on the other species can be written as β/K_2. β is the competition coefficient of the second population. Conversely, the effect of one individual of N_2 on the growth of species population N_1 is α/K_1, α being the competition coefficient of the first population. Population growth of the two populations can be represented by the following pair of equations:

$$\frac{dN_1}{dt} = r_1 N_1 \frac{K_1 - N_1 - \alpha N_2}{K_1}$$

$$\frac{dN_2}{dt} = r_2 N_2 \frac{K_2 - N_2 - \beta N_1}{K_2}$$

There are four possible outcomes of the competitive interaction and these are dependent on the following sets of inequalities:

1. When $\alpha < K_1/K_2$ and $\beta > K_2/K_1$, then N_1 is the winner. N_2 goes to extinction.
2. When $\alpha > K_1/K_2$ and $\beta < K_2/K_1$, then N_2 is the winner. N_1 goes to extinction.
3. When $\alpha > K_1/K_2$ and $\beta > K_2/K_1$, then either N_1 or N_2 alone will survive, depending on the initial densities of the two species.
4. When $\alpha < K_1/K_2$ and $\beta < K_2/K_1$, then the two species will coexist indefinitely since each inhibits its own growth more than that of the other species.

The seashore may or may not represent a smooth gradient of conditions, but in either case, the environment varies vertically. This means that the carrying capacity K for each species will vary with tidal height. When this happens there will be changes in the inequalities expressed above such that at a given level on the shore inequality number 1 may become switched to inequality number 2. For practical purposes this is an "all-or-nothing" or threshold event. As soon as the inequalities are changed there will be a change in the outcome of competitive interaction. This explanation does not depend on physiological tolerance limits of plants. Instead emphasis is placed on the carrying capacity of the environment for each species.

This model gives a conceptual basis to the observation that sharp interspecific lines of discontinuity can be caused by competition. Whether or not the model is valid is another matter and must await experimental verification. The ideas expressed above were developed by Pielou (1974), who gives alternative competition models.

SUCCESSION: STRUCTURE IN TIME

In the North Atlantic and in some other temperate waters the intertidal zone gives an impression of great stability through time. Lewis (1977) has noted that *Ascophyllum nodosum* "lives probably for several decades at least and forms a climax type vegetation which perennially determines the type and abundance of the accompanying species." This contrasts with the views of Connell and Slatyer (1977), who state that "we have found no example of a community of sexually reproducing individuals in which it has been demonstrated that the average species composition has reached a steady-state equilibrium. Until this is demonstrated, we conclude that, in general, succession never stops." How can these two conflicting viewpoints be resolved? It is hoped this will become clear in the following discussion.

In this treatment succession is regarded in the narrow sense of change in species composition through time after disturbance opens up a relatively large space for colonization. The creation of clear areas is a common natural phenomenon on rocky shores. The force of sea waves can easily shift boulders the size of grand pianos. Ice scouring also removes living organisms, as do biotic effects such as grazing. The normal sequence of colonization or succession then usually passes through the following stages: diatoms → fast growing algae and sessile animals → larger algae and animals. The species vary with habitat and geography, but the general sequence is usually the same. There are important exceptions however, and these are considered later. Traditionally it has been assumed that succession begins with colonization by pioneer species because these are the only forms that can exist immediately following perturbation. Subsequently the pioneer species are seen to modify the environment, making it more suitable for later species in the succession. This process continues until a climax species becomes dominant and arrests the succession. Connell and Slatyer (1977) term this conception of the succession process a facilitation mechanism. These authors propose two alternative models, which are shown along with the facilitation scheme in Figure 65. Sequence 1 is the facilitation model. Model 2 is the tolerance model. In this model later successional species are successful whether earlier species have preceded them or not. The later species can become established and grow to maturity in the presence of other species because they can grow at lower levels of resources than can earlier ones. Model 3 is the inhibition model. It hypothesizes that later species cannot grow to maturity in the presence of earlier colonists. These earlier colonists may be removed by natural mortality, from physical extremes, or the effects of herbivores. In the facilitation and tolerance models the colonists are killed in competition with the later species.

Evidence for the facilitation model is minimal for seaweed communities. Harger and Tustin (1973) suggest that the kelp *Ecklonia radiata* in New Zealand may colonize only after bryozoans have become established. Settling plates covered with ascidians were not colonized by *Ecklonia* and neither were plates free of sessile organisms. Plates covered with nylon fur were soon settled with *Ecklonia*. Most evidence from marine benthic environments supports model 3. A good example is the work of Dayton (1973). *Postelsia palmaeformis* is an annual kelp species which, on the west coast of North America, forms a strikingly contiguous nonoverlapping distribution with patches of the mussel *Mytilus californianus*. *Postelsia* sporophytes are annual whereas the mussel is long lived. Also, the mussel is known to be a competitive dominant and yet the kelp is able to maintain patches over time. Annual species normally constitute the early part of a succession. However, *Postelsia* is able to arrest the succession by smothering other organisms and/or causing them to be ripped from the substrate. In other words, the first colonizer holds the site against all comers, and this accords with Connell and Slatyer's (1977) inhibition model 3.

More evidence in favor of the inhibition model comes from studies by Norton and Burrows (1969). They found that when a large area is cleared in a forest of the long-lived kelp *Laminaria hyperborea* a dense stand of another kelp *Saccorhiza polyschides* develops. *Saccorhiza* easily holds its own against *L. hyperborea* through the summer. However the sporophytes of *Saccorhiza* are annual and disappear

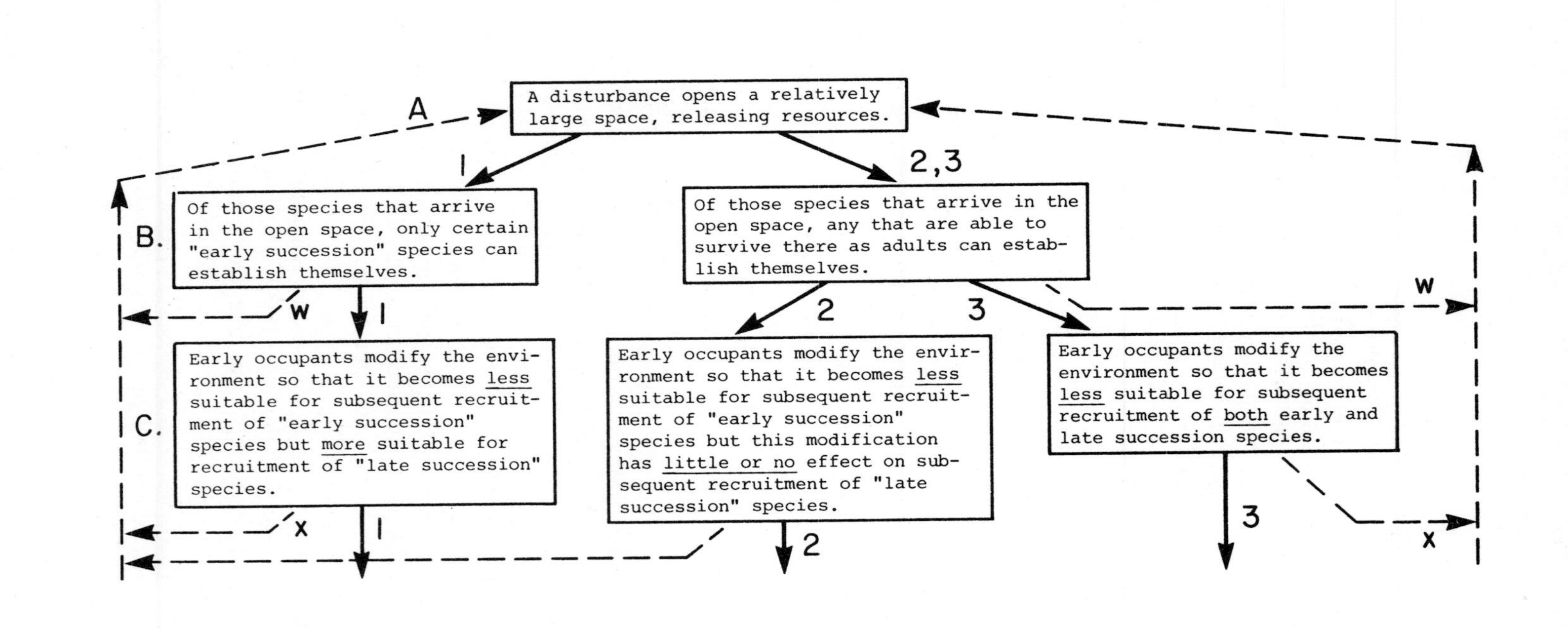
A
A disturbance opens a relatively large space, releasing resources.
1
2,3
B.
Of those species that arrive in the open space, only certain "early succession" species can establish themselves.
Of those species that arrive in the open space, any that are able to survive there as adults can establish themselves.
w
1
2
3
w
C.
Early occupants modify the environment so that it becomes less suitable for subsequent recruitment of "early succession" species but more suitable for recruitment of "late succession" species.
Early occupants modify the environment so that it becomes less suitable for subsequent recruitment of "early succession" species but this modification has little or no effect on subsequent recruitment of "late succession" species.
Early occupants modify the environment so that it becomes less suitable for subsequent recruitment of both early and late succession species.
x
1
2
3
x

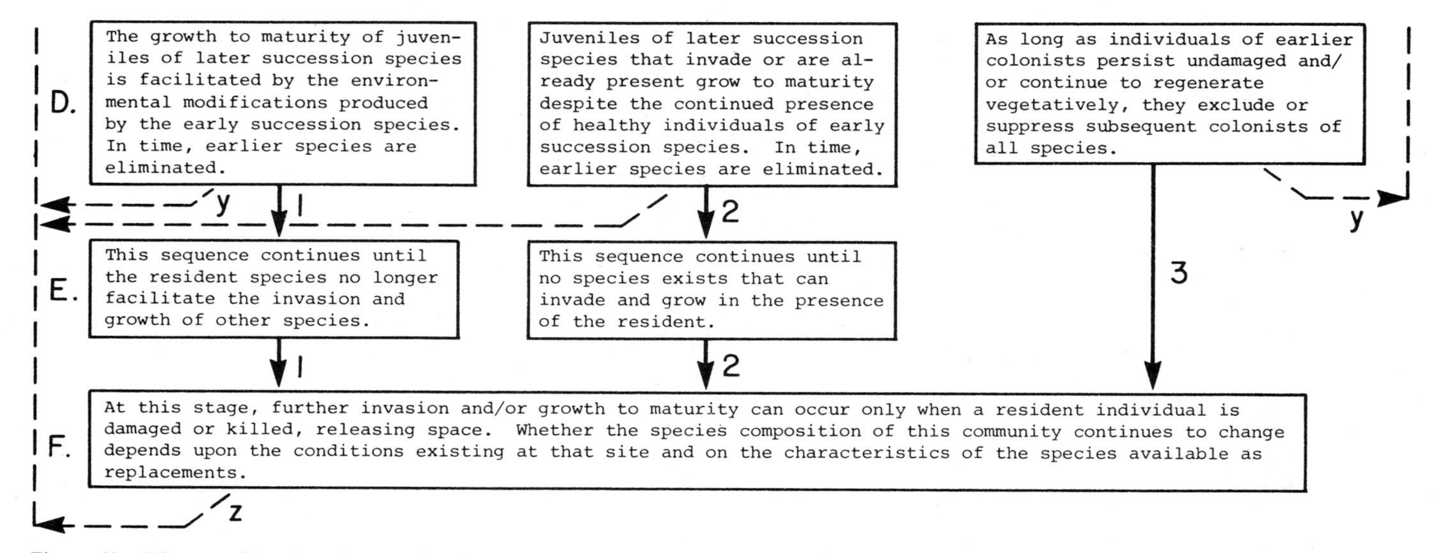

Figure 65. Three models of mechanisms producing species succession. The dashed lines represent interruptions of the succession process, in decreasing frequency in the order w, x, y, and z. See text for full explanation. From Connell and Slatyer (1977), *American Naturalist*. Copyright © 1977 by the University of Chicago.

during the winter. As soon as the space occupied by the *Saccorhiza* canopy was cleared, young sporophytes of *L. hyperborea* (presumably held in a state of suspended development in the deep shade), began to grow and occupy the site for perhaps 15 years. According to this view of succession, "climax" species are simply those that are longest lived. The stability of the climax depends upon whether individuals are replaced by members of their own or of other species. If the former, then there is species stability. We have already seen that experiments by Norton and Burrows demonstrate that in *L. hyperborea* forests replacement involves another species, *S polyschides*. In Europe the mid-intertidal region is commonly covered by dense stands of the perennial species *Ascophyllum nodosum*, and Lewis (1977) has referred to the long-term community stability imposed by this species. In fact, evidence by Burrows and Lodge (1951) appears to contradict this viewpoint. When *Ascophyllum* was cleared or cut, species of *Fucus*, present as sporelings beneath the canopy, began a phase of rapid growth and, in the case of *F. spiralis*, plants fruited the following year. In other words species stability is illusive. Succession never really comes to an end on seashores. The experiments reported by Burrows and Lodge (1951) support the earlier work of Pyefinch (1943). Pyefinch was surprised that small clearings in the *Ascophyllum* zone were soon colonized by *Fucus* species. It is hoped that, more than 30 years later, we no longer find this surprising.

References

Baker, J. R. J., and Evans, L. V. 1973. The ship fouling alga *Ectocarpus*. I. Ultrastructure and cytochemistry of plurilocular reproductive stages. Protoplasma 77:1–13.

Bentrup, F. W. 1963. Vergleichende Untersuchungen zur Polaritätsinduktion durch das Licht an der *Equisetum*—spore und der *Fucus*—zygote. Planta 59:472–491.

Berthold, G. 1882. Beiträge zur Morphologie und Physiologie der Meeresalgen. Jahr. Wiss. Bot. 13:569–717.

Bisalputra, T. 1974. Plastids. In: W. D. P. Stewart (ed.), Algal Physiology and Biochemistry, pp. 124–160. Blackwell Scientific Publications Ltd., Oxford.

Bold, H. C. 1967. Morphology of Plants. Harper and Row Publishers, New York.

Bold, H. C., and Wynne, M. 1978. Introduction to the Algae: Structure and Reproduction. Prentice Hall, Englewood Cliffs.

Boney, A. D. 1965. Aspects of the biology of the seaweeds of economic importance. Adv. Mar. Biol. 3:105–253.

Boney, A. D. 1966. A Biology of Marine Algae. Hutchinson & Co., London.

Bouck, G. B. 1965. Fine structure and organelle associations in brown algae. J. Cell Biol. 26:523–537.

Bouck, G. B. 1969. Extracellular microtubules: The origin, structure, and attachment of flagellar hairs in *Fucus* and *Ascophyllum* antherozoids. J. Cell Biol. 40:446–460.

Boudouresque, C.-F. 1970. Recherches sur les concepts de biocoenose et de continuum au niveau de peuplements benthiques sciaphiles. Vie et Milieu 21:103–136.

Boudouresque, C.-F. 1971. Recherches de bionomie analatique structurale et expérimentale sur les peuplements benthiques schiaphile de Méditerranée Occidentale (Fraction algale): La sous-strate sciaphile des peuplements de grandes *Cystoseira* de mode battu. Bull. Mus. Hist. Nat. Marseille 31:141–151.

Boudouresque, C.-F., and Lück, H. B. 1972. Recherches de bionomie structurale au niveau d'un peuplement benthique sciaphile. J. Exp. Mar. Biol. Ecol. 8:133–144.

Brachet, J. 1957. Biochemical Cytology. Academic Press, New York.

Bråten, T., and Nordby, Ø. 1973. Ultrastructure of meiosis and centriole behaviour in *Ulva mutabilis* Føyn. J. Cell. Sci. 13:69–81.

Buggeln, R. G. 1976. Auxin, an endogenous regulator of growth in algae? J. Phycol. 12:355–358.

Buggeln, R. G., and Craigie, J. S. 1971. Evaluation of evidence for the presence of indole-3-acetic acid in marine algae. Planta 97:173–178.

Burrows, E. M., and Lodge, S. 1951. Autecology and the species problem in *Fucus*. J. Mar. Biol. Assoc. (U.K.) 30:161–175.

Chamberlain, A. H. L., and Evans, L. V. 1973. Aspects of spore production in the red alga *Ceramium*. Protoplasma 76:139–159.

Chapman, A. R. O., and Craigie, J. S. 1977. Seasonal growth in *Laminaria longicruris*: Relations with dissolved inorganic nutrients and internal reserves of nitrogen. Mar. Biol. 40:197–205.

Chapman, V. J. 1970. Seaweeds and Their Uses. Methuen & Co. Ltd., London.

Charters, A. C., Neushul, M., and Barilotti, C. 1969. The functional morphology of *Eisenia arborea*. In: R. Margalef (ed.), International Seaweed Symposium, 6th, Santiago de Compostela 1968. Proceedings, pp. 89–105. Divección General de Pesca Maritima, Madrid.

Cheney, D. P., and Babbel, G. R. 1975. Isoenzyme variation in Florida populations of the red alga *Eucheuma*. J. Phycol. 11 (suppl.):17–18.

Connell, J. H. 1961. The influence of interspecific competition and other factors on the distribution of the barnacle *Chthamalus stellatus*. Ecology 42:710–723.

Connell, J. H. 1972. Community interactions on marine rocky intertidal shores. Annu. Rev. Ecol. Syst. 3:169–192.

Connell, J. H., and Slatyer, R. O. 1977. Mechanisms of succession in natural communities and their role in community stability and organization. Am. Nat. 111:1119–1144.

Davson, H. 1970. A Textbook of General Physiology. Little, Brown & Co., Boston.

Dawes, C. J. 1971. Indole-3-acetic acid in the green algal coenocyte *Caulerpa prolifera*. (Chlorophyceae, Siphonales). Phycologia 10:375–379.

Dawson, E. Y. 1966. Marine Botany. An Introduction. Holt, Rinehart & Winston, New York.

Dayton, P. K. 1971. Competition, disturbance and community organization: The provision and subsequent utilization of space in a rocky intertidal community. Ecol. Monogr. 41:351–389.

Dayton, P. K. 1973. Dispersion, dispersal, and persistence of the annual intertidal alga, *Postelsia palmaeformis* Ruprecht. Ecology 54:433–438.

Dayton, P. K. 1975a. Experimental studies of algal canopy interactions in a sea-otter dominated kelp community at Amchitka Island, Alaska. Fish. Bull. 73:230–237.

Dayton, P. K. 1975b. Experimental evaluation of ecological dominance in a rocky intertidal algal community. Ecol. Monogr. 45:137–159.

den Hartog, C. 1959. The epilithic algal communities occurring along the coast of the Netherlands. Wentia 1:1–241.

den Hartog, C. 1968. The littoral environment of rocky shores as a border between the sea and the land, and between the sea and the fresh water. Blumea 16:374–393.

Dixon, P. S. 1971. Cell enlargement in relation to the development of thallus form in Florideophyceae. Br. Phycol. J. 6:195–205.

Dixon, P. S. 1973. Biology of the Rhodophyta. Oliver and Boyd, Edinburgh.
Dodge, J. D. 1973. The Fine Structure of Algal Cells. Academic Press, London.
Doty, M. S. 1946. Critical tide factors that are correlated with the vertical distribution of marine algae and other organisms along the Pacific coast. Ecology 27:315–328.
Doty, M. S. 1957. Rocky intertidal surfaces. Geol. Soc. Am. Mem. 67 (1):535–585.
Doty, M. S., and Archer, J. G. 1950. An experimental test of the tide factor hypothesis. Am. J. Bot. 37:458–464.
Drew, K. M. 1939. An investigation of *Plumaria elegans* (Bonnem.) Schmitz with special reference to triploid plants bearing parasporangia. Ann. Bot. (N. S.) 3:347–367.
Drew, K. M. 1955. Life histories in the algae with special reference to the Chlorophyta, Phaeophyta and Rhodophyta. Biol. Rev. 30:343–390.
Dring, M. J. 1967. Effects of daylength on growth and reproduction of the *Conchocelis*-phase of *Porphyra tenera.* J. Mar. Biol. Assoc. (U.K.) 47:501–510.
Dring, M. J. 1971. Light quality and the photomorphogensis of algae in marine environments. In: D. J. Crisp (ed.), European Marine Biology Symposium, 4th, Bangor, 1969, Proceedings, pp. 417–424. Cambridge University Press, Cambridge.
Dring, M. J. 1974. Reproduction. In: W. D. P. Stewart (ed.), Algal Physiology and Biochemistry, pp. 814–837. Blackwell Scientific Publications Ltd, Oxford.
Dring, M. J., and Lüning, K. 1975. A photoperiodic response mediated by blue light in the brown alga *Scytosiphon lomentaria.* Planta (Berlin) 125:25–32.
Droop, M. R. 1974. Heterotrophy of carbon. In: W. D. P. Stewart (ed.), Algal Physiology and Biochemistry, pp. 530–559. Blackwell Scientific Publications Ltd., Oxford.
Edwards, P., Bird, E., Cotgreave, B., Cossins, A., Crompton, K., Fowler, W., Herdson, D., and Hudson, J. 1975. Marine phytobenthos of the Castellabate (Cilento) Natural Park, Salerno, Italy. Phytocoenolgia 1:403–426.
Engelmann, T. W. 1884. Untersuchungen über die quantitativen Beziehungen zwischen Absorption des Lichtes und Assimilation der Planzenzellen. Bot. Zeit. 42:81–93, 97–105.
Eppley, R. W., and Cyrus, C. C. 1960. Cation regulation and survival of the red alga, *Porphyra perforata*, in diluted and concentrated sea water. Biol. Bull. 118:55–65.
Evans, L. V. 1965. Cytological studies in the Laminariales. Ann. Bot. (N. S.) 29:541–562.
Evans, L. V. 1974. Cytoplasmic organelles. In: W. D. P. Stewart (ed.), Algal Physiology and Biochemistry, pp. 86–123. Blackwell Scientific Publications Ltd., Oxford.
Evans, L. V., Simpson, M., and Callow, M. E. 1973. Sulphated polysaccharide synthesis in brown algae. Planta 110:237–252.
Fischer, E. 1929. Recherches de bionomie et d'océanographie littorales sur la Rance et le littoral de La Manche. Ann. Inst. Oceanogr. Monaco (N. S.) 5:205–429.

Floc'h, J.-Y, and Penot, M. 1972. Transport du ^{32}P et du ^{86}Rb chez quelques alques brunes: Orientation des migrations de voies de conduction. Physiol. Veg. 10:677–686.
Fritsch, F. E. 1935. The Structure and Reproduction of the Algae, Vol. I. Cambridge University Press, Cambridge.
Fritsch, F. E. 1945. Structure and Reproduction of the Algae. Vol. II. Cambridge University Press, Cambridge.
Gayral, P. 1975. Les Algues. Morphologie, Cytologie, Reproduction, Ecologie. Doin, Paris.
Gessner, F., and Schramm, W. 1971. Salinity. In: O. Kinne (ed.), Marine Ecology, Vol. I, Part 2, pp. 705–820. John Wiley & Sons, New York.
Giese, A. C. 1973. Cell Physiology. W. B. Saunders Co., Philadelphia.
Govindjee and Braun, B. Z. 1974. Light absorption, emission and photosynthesis. In: W. D. P. Stewart (ed.), Algal Physiology and Biochemistry, pp. 346–390. Blackwell Scientific Publications Ltd., Oxford.
Green, P. B. 1969. Cell morphogenesis. Annu. Rev. Plant Physiol. 20:365–394.
Green, P. B., Erickson, R. O., and Richmond, P. A. 1970. On the physical basis of cell morphogenesis. Ann. N.Y. Acad. Sci. 175:712–731.
Gutknecht, J., and Dainty, J. 1968. Ionic relations of marine algae. Oceanogr. Mar. Biol. Annu. Rev. 6:163–200.
Hall, I. 1975. A parallel study of the enzymes xanthine dehydrogenase and malic dehydrogenase in natural populations of Enteromorpha intestinalis. Bachelor of Science thesis. Dalhousie University, Halifax.
Hämmerling, J. 1963. Nucleo-cytoplasmic interactions in *Acetabularia* and other cells. Annu. Rev. Plant Physiol. 14:65–92.
Hanic, L. A., and Craigie, J. S. 1969. Studies on the algal cuticle. J. Phycol. 5:89–102.
Harger, J. R. E., and Tustin, K. 1973. Succession and stability in biological communities. I. Diversity. Int. J. Environment. Stud. 5:117–130.
Hartmann, M. 1925. Untersuchungen über relative sexüalitat. Biol. Zbl. 45:449–467.
Hellebust, J. A. 1976. Osmoregulation. Annu. Rev. Plant Physiol. 27:485–505.
Kain, J. M. 1971. Synopsis of biological data on *Laminaria hyperborea*. FAO Fish. Synopsis No. 87.
Kain, J. M. 1975. The biology of *Laminaria hyperborea*. VII. Reproduction of the sporophyte. J. Mar. Biol. Assoc. (U.K.) 55:567–582.
Kauss, H. 1968. α-Galaktosylglyzeride und Osmoregulation in Rotalgen. Z. Pflanzenphysiol. 58:428–433.
Klein, R. M., and Cronquist, A. 1967. A consideration of the evolutionary and taxonomic significance of some biochemical, micromorphological, and physiological characters in the Thallophytes. Quart. Rev. Biol. 42:105–296.
Köhler, K. 1956. Entwicklungsgeschichte, Geschlechtsbestimmung und Befruchtung bei *Chaetomorpha*. Arch. Protistenk. 101:223–268.
Krishnamurthy, V. 1959. Cytological investigations on *Porphyra umbilicalis* (L.) Kutz. var. *laciniata* (Lightf.) J. Ag. Ann. Bot. (N. S.) 23:147–176.
Kugrens, P., and West, J. A. 1972. Synaptonemal complexes in red algae. J. Phycol. 8:187–191.

Levring, T. 1947. Submarine daylight and the photosynthesis of marine algae. Göteborgs Vetensk Samh. Handl., IV Ser., B, 5/6:1–89.

Levring, T. 1966. Submarine light and algal shore zonation. In: R. Bainbridge, G. C. Evans, and O. Rackham (eds.), Light as an Ecological Factor, pp. 305–318. British Ecological Society Symposium Volume No. 6. Blackwell Scientific Publications Ltd., Oxford.

Lewis, J. R. 1961. The littoral zone on rocky shores—a biological or physical entity? Oikos 12:280–301.

Lewis, J. R. 1964. The Ecology of Rocky Shores. English Universities Press Ltd., London.

Lewis, J. R. 1977. The role of physical and biological factors in the distribution and stability of rocky shore communities. In: B. F. Keegan (ed.), European Marine Biology Symposium, 11th, Galway, 1976. Biology of Benthic Organisms, pp. 417–424. Pergamon, New York.

Lichtenthaler, H. K. 1968. Plastoglobuli and the fine structure of plastids. Endeavor 27:144–149.

Løvlie, A., and Bråten, T. 1970. On mitosis in the multicellular alga *Ulva mutabilis*. Føyn. J. Cell. Sci. 6:109–129.

Lüning, K., and Dring, M. 1975. Reproduction, growth and photosynthesis of gametophytes of *Laminaria saccharina* grown in blue and red light. Mar. Biol. 29:195–200.

Lüning, K., Schmitz, K., and Willenbrink, J., 1973. CO_2 fixation and translocation in benthic marine algae. III. Rates and ecological significance of translocation in *Laminaria hyperborea* and *L. saccharina*. Mar. Biol. 23:275–281.

McDonald, K. 1972. The ultrastructure of mitosis in the marine red alga *Membranoptera platyphylla*. J. Phycol. 8:156–166.

Mackie, W., and Preston, R. D. 1974. Cell wall and intercellular region polysaccharides. In: W. D. P. Stewart (ed.), Algal Physiology and Biochemistry, pp. 40–85. Blackwell Scientific Publications Ltd., Oxford.

Manton, I. 1964. A contribution towards understanding of "The Primitive Fucoid." New Phytol. 63:244–254.

Manton, I., and Clarke, B. 1951. An electron microscope study of the spermatozoid of *Fucus serratus*. Ann. Bot. (N. S.) 15:461–471.

Menge, B. A. 1976. Organization of the New England rocky intertidal community: Role of predation, competition and environmental heterogeneity. Ecol. Monogr. 46:355–393.

Mills, E. L. 1969. The community concept in marine zoology, with comments on continua and instability in some marine communities: A review. J. Fish. Res. Board Can. 26:1415–1428.

Moss, B. 1967. The apical meristem of *Fucus*. New Phytol. 66:67–74.

Moss, B. 1974. Morphogenesis. In: W. D. P. Stewart (ed.), Algal Physiology and Biochemistry, pp. 788–813. Blackwell Scientific Publications Ltd., Oxford.

Müller, D. G. 1962. Über jahres-und lunarperiodische Ercheinungen bei einigen Braunalgen. Bot. Mar. 4:140–155.

Müller, D. G. 1967. Generationswechsel, Kernphasenwechsel und Sexualität der Braunalge *Ectocarpus siliculosus* im Kulturvensuch. Planta (Berlin) 75:39–54.

Müller, D. G. 1968. Versuche zur Charakterisierung eines Sexual-Lockstoffes bei der Braunalge *Ectocarpus siliculosus*. Planta (Berlin) 81:160–168.
Müller, D. G. 1976a. Relative sexuality in *Ectocarpus siliculosus* (Phaeophyta). A scientific error. Arch. Microbiol. 109:89–94.
Müller, D. G. 1976b. Sexual isolation between a European and an American population of *Ectocarpus siliculosus*. J. Phycol. 12:252–254.
Müller, D. G., and Jaenicke, L. 1973. Fucoserraten, the female sex attractant of *Fucus serratus*, L. (Phaeophyta). FEBS Letters 30:137–139.
Müller, D. G., Jaenicke, L., Donike, M., and Akintobi, T. 1971. Sex attractant in a brown alga: Chemical structure. Science 171:815–817.
Nakahara, H., and Tatewaki, M. 1971. Some differences in nutritional requirements between different generations in the brown alga *Desmarestia*. Bot. Mag. (Tokyo) 84:435–437.
Nakazawa, S. 1975. Physiology of *Fucus*. In: J. Tokida and H. Hirose (eds.), Advance of Phycology in Japan, pp. 160–170. Junk, The Hague.
Neushul, M. 1972. Functional interpretation of benthic marine algal morphology. In: I. A. Abbott and M. Kurogi (eds.), Contributions to the Systematics of Benthic Marine Algae of the North Pacific, pp. 47–73, Japanese Society of Phycology, Kobe.
Nicholson, N. L. 1976. Anatomy of the medulla of *Nereocystis*. Bot. Mar. 19:23–31.
Norton, T. A., and Burrows, E. M. 1969. Studies on marine algae of the British Isles. 7. *Saccorhiza polyschides* (Lightf.) Batt. Br. Phycol. J. 4:19–53.
Norton, T. A., McAllister, H. A., and Conway, E. 1969. The marine algae of the Hebridean island of Colonsay. Br. Phycol. J. 4:125–136.
Nultsch, W. 1974. Movements. In: W. D. P. Stewart (ed.), Algal Physiology and Biochemistry, pp. 864–893. Blackwell Scientific Publications Ltd., Oxford.
O'Kelley, J. C. 1974. Inorganic nutrients. In: W. D. P. Stewart (ed.), Algal Physiology and Biochemistry, pp. 610–635. Blackwell Scientific Publications Ltd., Oxford.
Page, J. Z., and Sweeney, B. M. 1968. Culture studies on the marine green alga *Halicystis parvula-Derbesia tenuissima*. III. Control of gamete formation by an endogenous rhythm. J. Phycol. 4:253–260.
Paine, R. T. 1971. A short-term experimental investigation of resource partitioning in a New Zealand rocky intertidal habitat. Ecology 52:1096–1106.
Paine, R. T. 1974. Intertidal community structure: Experimental studies on the relationship between a dominant competitor and its principal predator. Oecologia 15:93–120.
Pearl, R. 1927. The growth of populations. Quart. Rev. Biol. 2:532–548.
Percival, E., and McDowell, R. H. 1967. Chemistry and Enzymology of Marine Algal Polysaccharides. Academic Press, New York.
Pickett-Heaps, J. D. 1975. Green Algae, Structure, Reproduction and Evolution in Selected Genera. Sinauer Associates, Sunderland.
Pielou, E. C. 1974. Competition on an environmental gradient. In: P. van den Driessche (ed.), Proc. Conf. Math. Probl. Biol.: 184–204. Springer-Verlag, Berlin.

Prentice, S. A., and Kain, J. M. 1976. Numerical analysis of subtidal communities on rocky shores. Est. Coast. Mar. Sci. 4:65–70.

Prescott, G. W. 1968. The Algae: A review. Houghton Mifflin Co., Boston.

Preston, R. D. 1974. The Physical Biology of Plant Cell Walls. Chapman and Hall, London.

Provasoli, L., and Carlucci, A. F. 1974. Vitamins and growth regulators. In: W. D. P. Stewart (ed.), Algal Physiology and Biochemistry, pp. 741–787. Blackwell Scientific Publications Ltd., Oxford.

Provasoli, L., and Pintner, I. J. 1964. Symbiotic relationships between microorganisms and seaweeds. (Abstr.) Am. J. Bot. 51:681.

Pyefinch, K. A. 1943. The intertidal ecology of Bardsey Island, North Wales, with special reference to the recolonization of rock surfaces, and the rock pool environment. J. Animal Ecol. 12:82–108.

Ramus, J. 1972. Differentiation of the green alga *Codium fragile*. Am. J. Bot. 59:478–482.

Ramus, J., Beale, S. I., and Mauzerall, D. 1976a. Correlation of changes in pigment content with photosynthetic capacity of seaweeds as a function of water depth. Mar. Biol. 37:231–238.

Ramus, J., Beale, S. I., Mauzerall, D., and Howard, K. L. 1976b. Changes in photosynthetic pigment concentration in seaweeds as a function of water depth. Mar. Biol. 37:223–229.

Raven, J. A. 1974. Carbon dioxide fixation. In: W. D. P. Stewart (ed.), Algal Physiology and Biochemistry, pp. 434–455. Blackwell Scientific Publications Ltd., Oxford.

Richardson, N., and Dixon, P. S. 1968. Life history of *Bangia fuscopurpurea* (Dillw.) Lyngb. in culture. Nature 218:496–497.

Ringo, D. L. 1967. Flagellar motion and fine structure of the flagellar apparatus in *Chlamydomonas*. J. Cell Biol. 33:543–571.

Robinson, D. G., and Preston, R. D. 1971. The fine structure of swarmers of *Cladophora* and *Chaetomorpha*. I. The plasmalemma and Golgi apparatus in naked swarmers. J. Cell Sci. 9:581–601.

Robinson, D. G., White, R. K., and Preston, R. D. 1972. Fine structure of swarmers of Cladophora and Chaetomorpha. III. Wall synthesis and development. Planta (Berlin), 107:131–144.

Rosenthal, R. J., Clarke, W. D., and Dayton, P. K. 1974. Ecology and natural history of a stand of giant kelp, *Macrocystis pyrifera*, off Del Mar, California. Fish Bull. 72:670–684.

Round, F. E. 1973. The Biology of the Algae. Edward Arnold, London.

Russell, G. 1964. *Laminariocolax tomentosoides* on the Isle of Man. J. Mar. Biol. Assoc. (U.K.) 44:601–612.

Russell, G. 1967. The genus *Ectocarpus* in Britain. II. The free-living forms. J. Mar. Biol. Assoc. (U.K.) 47:233–250.

Russell, G. 1972. Phytosociological studies on a two-zone shore. I. Basic pattern. J. Ecol. 60:539–545.

Russell, G. 1973. The "litus-line": A re-assessment. Oikos 24:158–161.

Russell, G., and Bolton, J. J. 1975. Euryhaline ecotypes of *Ectocarpus siliculosus* (Dillw.) Lyngb. Est. Coast Mar. Sci. 3:91–94.

Sachs, J. von 1882. Textbook of Botany. Clarendon Press, Oxford.

Sagromsky, H. 1961. Durch Licht-Dunkel-Wechsel induzierter Rhythmus der Entleerung der Tetrasporangien von *Nitophyllum punctatum*. Pubbl. Staz. Zool. Napoli 32:29–40.

Saito, Y., and Atobe, S. 1970. Phytosociological study of intertidal marine algae. I. Usujiri Benten-Jima, Hokkaido. Bull. Fac. Fish. (Hokkaido Univ.) 21:37–69.

Saito, Y., Taniguchi, K., Atobe, S., and Naganawa, S. 1971. Phytosociological study of the intertidal marine algae. II. The algal communities on the vertical substratum faces on several directions. Jpn. J. Ecol. 20:230–232.

Scagel, R. F. 1967. Guide to Common Seaweeds of British Columbia. British Columbia Provincial Museum, Handbook No. 27. Queen's Printer, Victoria, B.C.

Scagel, R. F., Bandoni, R. J., Rouse, G. E., Schofield, W. D., Stein, J. R., Taylor, T. M. C. 1965. An Evolutionary Survey of the Plant Kingdom. Wadsworth, Belmont.

Schmitz, K., and Lobban, C. S. 1976. A survey of translocation in Laminariales (Phaeophyceae). Mar. Biol. 36:207–216.

Schmitz, K., Lüning, K., and Willenbrink, J. 1972. CO_2 fixierung und Stofftransport in benthischen marinen Algen. II. Zum Ferntransport ^{14}C-markierter Assimilate bei *Laminaria hyperborea* und *Laminaria saccharina*. Z. Pflanzenphysiol. 67:418–429.

Schmitz, K., and Srivastava, L. M. 1975. On the fine structure of sieve tubes and the physiology of assimilate transport in *Alaria marginata*. Can. J. Bot. 53:861–876.

Schopf, J. W. 1970. Precambrian micro-organisms and evolutionary events prior to the origin of vascular plants. Biol. Rev. 45:319–352.

Simon-Bichard-Bréaud, J. 1971. Un appareil cinétique dans les gamétocystes mâles d'une Rhodophycée: *Bonnemaisonia hamifera*, Hariot. C. R. Hebd. Séanc. Acad. Sci. Paris, sér D., 273:1272–1275.

Sjöstedt, L. T. 1928. Littoral and supralittoral studies on the Scanian shores. Lunds. Univ. Årrskr. 24:1–36.

Smith, G. M. 1944. Marine Algae of the Monterey Peninsula, California. Stanford University Press, Stanford.

Steinbiss, H. H., and Schmitz, K. 1973. CO_2-Fixierung und stofftransport in benthischen marinen Algen. V. Zur autoradiographischen Lokalisation der Assimilattransportbahnen im Thallus von *Laminaria hyperborea*. Planta (Berlin) 112:253–263.

Stephenson, T. A., and Stephenson, A. 1949. The universal features of zonation between tide marks on rocky coasts. J. Ecol. 37:289–305.

Stephenson, T. A., and Stephenson, A. 1972. Life Between Tidemarks on Rocky Shores. W. H. Freeman & Co., San Francisco.

Stewart, K. D., and Mattox, K. R. 1975. Comparative cytology, evolution and classification of the green algae, with some consideration of the origin of other organisms with chlorophylls a and b. Bot. Rev. 41:104–135.

Subbaramaiah, K. 1970. Growth and reproduction of *Ulva fasciata* Delile in nature and in culture. Bot. Mar. 13:25–27.

Sweeney, B. M., and Hastings, J. W. 1962. Rhythms. In: R. A. Lewin (ed.), Physiology and Biochemistry of Algae, pp. 687–700. Academic Press, New York.

Taniguti, M. 1962. Phytosociological Study of Marine Algae in Japan. Inoue and Co., Tokyo.
Terborgh, J. 1965. Effects of red and blue light on the growth and morphogenesis of *Acetabularia crenulata*. Nature 207:1360–1363.
Toth, R. T. 1976. The release, settlement and germination of zoospores in *Chorda tomentosa* (Phaeophyceae, Laminariales). J. Phycol. 12:222–233.
Townsend, C., and Lawson, G. W. 1972. Preliminary results on factors causing zonation in *Enteromorpha* using a tide simulating apparatus. J. Exp. Mar. Biol. Ecol. 8:265–276.
Van den Hoek, C., Cortel-Breeman, A. M., and Wanders, J. B. W. 1975. Algal zonation in the fringing coral reef of Curaçao, Netherlands Antilles, in relation to zonation of Corals and Gorgonians. Aquatic Bot. 1:269–308.
Venkataraman, G. S., Goyal, S. K., Kaushik, B. D., and Roychoudhury, P. 1974. Algae: Form and Function. Today and Tomorrow's Printers and Publishers, New Delhi.
Whittaker, R. H. 1962. Classification of natural communities. Bot. Rev. 28:1–239.
Wynne, M. J., and Loiseaux, S. 1976. Recent advances in life history studies of the Phaeophyta. Phycologia 15:435–452.
Zaneveld, J. S. 1969. Factors controlling the delimitation of littoral benthic marine algal zonation. Am. Zool. 9:367–391.

Index